Transcriptomics in Entomological Research

Transcriptomics in Entomological Research

Edited by

Matan Shelomi

National Taiwan University

CABI

CABI is a trading name of CAB International

CABI
Nosworthy Way
Wallingford
Oxfordshire OX10 8DE
UK
Tel: +44 (0)1491 832111
Fax: +44 (0)1491 833508
E-mail: info@cabi.org
Website: www.cabi.org

CABI
745 Atlantic Avenue
8th Floor
Boston, MA 02111
USA
T: +1 (617)682-9015
E-mail: cabi-nao@cabi.org

A catalogue record for this book is available from the British Library, London, UK.

Library of Congress Cataloging-in-Publication Data

Names: Shelomi, Matan, editor.
Title: Transcriptomics in entomological research / editor: Matan Shelomi.
Description: Boston, MA : CAB International, [2020]. | Includes bibliographical references and index. | Summary: "While "genomics" is becoming a household term, knowing the genome of an organism alone provides relatively little information. Between the genome and the final organism is the "transcriptome" that tells us which genes are expressed and translated into proteins in a certain cell or tissue of an organism at a given time and specific situation. The transcriptome tells you what genes are being "transcribed" at any given moment, providing the wealth of data and specificity of proteomics with the relative ease of study of genetics. Transcriptomics technology thus has myriad uses, and the goal of this book is to showcase the extraordinary diversity of ways transcriptomics can be utilised in entomological research, from basic to applied, from ecology to physiology to agricultural"-- Provided by publisher.
Identifiers: LCCN 2019026935 (print) | LCCN 2019026936 (ebook) | ISBN 9781789243130 (hardback) | ISBN 9781789243147 (ebook) | ISBN 9781789243154 (epub)
Subjects: LCSH: Genetic transcription. | Insects--Genetics.
Classification: LCC QH450.2 .T6983 2020 (print) | LCC QH450.2 (ebook) | DDC 572.8/845--dc23
LC record available at https://lccn.loc.gov/2019026935
LC ebook record available at https://lccn.loc.gov/2019026936

ISBN-13: (paperback) 9781789243130
(ePDF) 9781789243147
(ePub) 9781789243154

Commissioning Editor: Ward Cooper
Editorial Assistant: Lauren Davies
Production Editor: James Bishop

Typeset by Exeter Premedia Services Pvt Ltd, Chennai, India
Printed and bound in the UK by Severn, Gloucester

Contents

Contributors

Berenbaum, May R.
University of Illinois, Urbana-Champaign
505 S. Goodwin Ave
Urbana, IL 61801, USA
maybe@illinois.edu

Calla, Bernarda
University of Illinois, Urbana-Champaign
505 S. Goodwin Ave
Urbana, IL 61801, USA
calla2@illinois.edu

Chiu, Joanna C.
University of California, Davis
1 Shields Ave
Davis, CA 95616, USA
jcchiu@ucdavis.edu

Ehlting, Jürgen
University of Victoria, Canada
Cunningham 202
3800 Finnerty Road
Victoria, BC V8P 5C2, Canada
je@uvic.ca

Gee, Melanie
University of California, Berkeley
Department of Environmental Science, Policy, and Management
130 Mulford Hall, #3114

Berkeley, CA 94720-3114, USA
melaniegee@berkeley.edu

Gill, Aman
University of California, Berkeley
Department of Environmental Science, Policy, and Management
130 Mulford Hall, #3114
Berkeley, CA 94720-3114, USA
amango@gmail.com

Huff, Matthew
University of Tennessee, Knoxville
153 Plant Biotechnology Building
2505 EJ Chapman Drive
Knoxville, TN 37996-4560
mhuff10@utk.edu

Johnson, Brian R.
University of California, Davis
1 Shields Ave
Davis, CA 95616, USA
brnjohnson@ucdavis.edu

Jurat-Fuentes, Juan Luis
University of Tennessee, Knoxville
2505 E. J. Chapman Drive
Knoxville, TN, 37996, USA
jurat@utk.edu

Klingeman, William E.
University of Tennessee, Knoxville
2431 Joe Johnson Drive
Knoxville, TN 37996-4561, USA
wklingem@tennessee.edu

Lewald, Kyle M.
University of California, Davis
1 Shields Ave
Davis, CA 95616, USA
kmlewald@ucdavis.edu

Liu, Shanlin
BGI-Shenzhen, China
Building 11, Beishan Industrial Zone, Yantian District
Shenzhen 518083, China
shanlin1115@gmail.com

Malacrinò, Antonino
Department of Evolution, Ecology and Organismal Biology
The Ohio State University
Columbus, OH, 43210, USA
antonino.malacrino@gmail.com, malacrino.1@osu.edu

Meng, Guanliang
BGI-Shenzhen, China
Building 11, Beishan Industrial Zone, Yantian District
Shenzhen 518083, China
mengguanliang2012@gmail.com

Paulson, Amber Rose
University of Victoria, Canada
Cunningham 202
3800 Finnerty Road
Victoria, BC V8P 5C2, Canada
amber.rose.paulson@gmail.com

Perlman, Steven J.
University of Victoria, Canada
Cunningham 202
3800 Finnerty Road
Victoria, BC V8P 5C2, Canada
stevep@uvic.ca

Pothula, Ratnasri
University of Tennessee, Knoxville
2505 E. J. Chapman Drive
Knoxville, TN, 37996, USA
rmallipe@vols.utk.edu

Sattar, Sampurna
Pennsylvania State University
University Park, PA 16082, USA
sus56@psu.edu

Shelomi, Matan
National Taiwan University
No 27 Lane 113 Sec 4 Roosevelt Rd
Taipei 10617, Taiwan
mshelomi@ntu.edu.tw

Staton, Margaret E
University of Tennessee, Knoxville
2505 E. J. Chapman Drive

Knoxville, TN, 37996, USA
mstaton1@utk.edu

Tauber, James P.
Beltsville Agricultural Research Center, USDA
10300 Baltimore Ave
Beltsville, MD 20705, USA
james.tauber@usda.gov, jamesptauber@gmail.com

Thompson, Gary A.
Pennsylvania State University
217 Ag Admin Building
University Park, PA 16802, USA
gat10@psu.edu

von Aderkas, Patrick
University of Victoria, Canada
Cunningham 202
3800 Finnerty Road
Victoria, BC V8P 5C2, Canada
pvonader@uvic.ca

Will, Kipling
Essig Museum of Entomology
1101 Valley Life Sciences Building, #4780
University of California, Berkeley
Berkeley, CA 94720-4780
kipwill@berkeley.edu

Zhou, Chengran
BGI-Shenzhen, China
Building 11, Beishan Industrial Zone, Yantian District
Shenzhen 518083, China
zhouchengran@genomics.cn

Preface by the Editor

Matan Shelomi

National Taiwan University, Taipei, Taiwan

The genome has been compared with a cookbook. Each gene is a recipe, with detailed instructions on which ingredients (amino acids) for the cooks (ribosomes) to combine in which order to produce the final dish (protein), with various non-coding regions that indicate what to cook when and where and even to generate the cooks themselves (ok, the metaphor breaks apart there). Each cell in an organism has the same genome, the same set of recipes to choose from to make the cells, their components, and their secretions. Just as a cookbook has appetizers, soups, mains, and desserts, so too does the genome cover far more proteins than any given cell needs to make at any given time.

If the genome is a cookbook, then the transcriptome is the order tickets: what that particular kitchen is in the process of making right then and there. The transcriptome is the messenger (RNA) between the genome and the ribosome, that says what protein to make right now at this place and time. A transcriptome is a snapshot of what proteins a cell is in the process of making: what genes are being transcribed for the ribosomes to translate. While a genome tells you what the organism can do in theory, a transcriptome tells you what the organism, tissue, or even single cell actually does and was doing at the time one extracted the RNA from it. By comparing different tissues or species, and/or varying the conditions or times or environments, one can see how the variables affect gene translation at that time, and help translate the wealth of potential information encoded in a genome into more practical and grounded data with tangible significance. Transcriptomics, the sequencing and analysis of transcriptomes, combines the relative simplicity of genomics with the empirical nature of proteomics.

Entomologists have been using transcriptomics for decades, but no reference on the subject of transcriptomics in the field had existed. It was each individual scientist's unspoken responsibility to discover transcriptomics on their own and decide whether and how they could use it, which of course means one usually will not encounter transcriptomics unless working with a senior researcher who already has. To bring such researchers and the curious neophytes together, in 2017 I organized the symposium 'Revelations from Insect Transcriptomics' at the annual meeting of the Entomological Society of America in Denver, Colorado, USA. My goals were to showcase not only the diversity of ways in which transcriptomics can be used within entomology, but also the diversity of the scientists themselves. The invited speakers included students and tenured faculty alike, consisted predominantly of women's voices, and represented a broad range of races, nationalities, sexualities, and abilities. The event drew a larger crowd than anticipated, and caught the eye of the publishers. The book you are reading now is the direct product of this symposium, and many of its chapter authors were original symposium speakers.

The purpose of this book is to serve as an introduction to transcriptomics and present an array of its uses in entomology, past and present, though of course it can be just as easily applied to any other branch of biology. I have laid out the menu from general to specific. We start with a thorough introduction to how transcriptomics works, its history, and some of its broad-stroke uses in insect science (Chapter 1). This chapter is followed by an exhaustively comprehensive look at how transcriptomics analysis is performed, and the software packages available to help one go from next-generation sequencing data to an interpretable dataset (Chapter 2).

We continue with reviews of how transcriptomics has been used within larger subfields of entomology, showing different applications of the basic techniques covered earlier. Transcriptomics and other next-generation sequencing technologies are ushering in radical new ways to approach pest management by finding new targets for control, genes responsible for pesticide resistance, and even novel pathogens (Chapter 3). This is exemplified by the many studies done on the aphids, often with non-model but economically significant species for whom genomic data does not exist, which have succeeded in finding critical genes involved in aphid–plant interactions and host specificity and finding targets for biocontrol by blocking transcription of key genes (Chapter 4). Identifying new pathogens has been particularly important for honey bees, where transcriptomics revealed several new pathogens with potential links to Colony Collapse Disorder (Chapter 5). Discoveries within insects can have implications throughout biology: researchers use transcriptomics to accelerate the discovery of certain large yet conserved gene families, such as the hyper-diverse and multifunctional cytochrome P450s of interest to toxicologists, physiologists, agriculturalists, pharmacologists, and more (Chapter 6).

We end with case studies going into more depth on how transcriptomics has been used to reveal more specific facets of a particular system. With its high power and low bias, transcriptomics is ideally suited to generate information for non-model organisms, cryptic species, and organisms that cannot be cultured in a laboratory, including insects, endocellular symbiotic microbes, or even both at the same time. The cases presented here include untangling insect–microbe interactions in cryptic parasitoid wasps (Chapter 7), identifying conserved insect digestive enzymes from the silverfish transcriptome (Chapter 8), discovering the function of mysterious organs in the stick insects (Chapter 9), and using functional transcriptomics to describe the chemical defenses of the bombardier beetle (Chapter 10).

It is my hope that the readers of this book will be inspired by the many possibilities transcriptomics offers to find a way to apply it to their needs and their research systems, and that the chapters herein can provide some practical information on how to get started. The power of this method to reveal the unknown is immense, and we have barely scratched the surface. Do not take my word for it, though. The authors and their references can speak for themselves.

Special thanks to Ward Cooper for suggesting this book, the Entomological Society of America for bringing all relevant minds together, and the contributors for donating their time and text to this tome.

1 Harnessing Transcriptomics to Study Insect Biology

KYLE M. LEWALD,* AND JOANNA C. CHIU

Department of Entomology and Nematology, University of California Davis, USA

1.1 Introduction

Over the past decade, the development and democratization of high-throughput sequencing has pushed biological investigations into a new era of massive data collection and analysis, unparalleled by anything seen before. With published genomes becoming a standard step for studying model and non-model organisms alike, and with large collaborative projects such as the i5k Insect Genome Project (i5k Consortium, 2013; Thomas *et al.*, 2018) and Earth BioGenome Project (Lewin *et al.*, 2018) underway, genomic data will only become increasingly accessible and valuable for comparative studies. Despite the broad utility of genome sequences for investigating various aspects of biology (e.g. genome structure, heredity, evolutionary mechanisms), assessing the functional potential of a gene requires the study of transcriptomes, namely repertoires of transcripts expressed in specific spatial and temporal patterns in specific physiological conditions (Wang *et al.*, 2009). For example, alternative splicing allows for one gene to generate multiple isoforms, but which one is most relevant or prevalent, and in what conditions? Bioinformatic analysis can be leveraged to identify promoters, enhancers, and transcription factor binding sites to predict expression, but ultimately these models need to be verified experimentally.

The emerging field of transcriptomics can provide functional data that can answer questions far beyond those that can be answered through genomic analysis alone. Its broad utility allows its applications in topics as far ranging as gene regulation, development, evolution, environmental responses, toxicology, immunity, and host–parasite interactions. In this chapter, we discuss the history and development of transcriptomics (Fig. 1.1) and common

*Corresponding author: kmlewald@ucdavis.edu

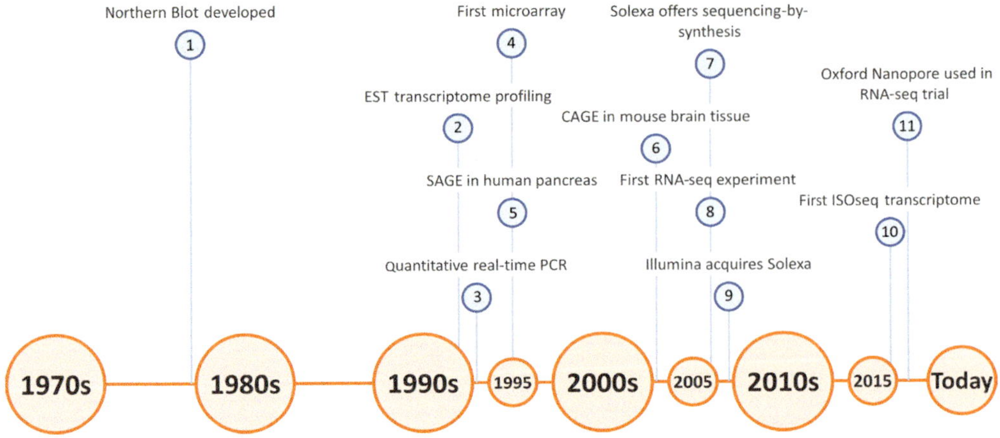

Fig. 1.1. Timeline for the development of transcriptomic technologies. X-axis indicates time and circular markers indicate notable experiments and events. Numbers within points denote source reference. (1) Alwine *et al.*, 1977 (2) Okubo *et al.*, 1992 (3) Higuchi *et al.*, 1993 (4) Schena *et al.*, 1995 (5) Velculescu *et al.*, 1995 (6) Shiraki *et al.*, 2003 (7) https://www.illumina.com/science/technology/next-generation-sequencing/illumina-sequencing-history.html (8) Lister *et al.*, 2008; Mortazavi *et al.*, 2008; Nagalakshmi *et al.*, 2008 (9) See reference 7. (10) Wang *et al.*, 2016. (11) Hargreaves and Mulley, 2015.

methodologies (Table 1.1), and highlight case studies of transcriptomics in entomology.

1.2 Technologies to Assay Gene Expression

1.2.1 Pre-transcriptomics

Prior to the development of modern genomic technologies, researchers were mostly limited to analyzing transcripts of a few genes at a time. One of the earliest methods of studying RNA transcripts was through the use of Northern blotting, developed as a variation of the Southern blot in the 1970s (Alwine *et al.*, 1977). In this procedure, RNA extracted from a sample is size-separated using gel electrophoresis, then transferred onto a nylon membrane. Radioactive probes complementary to the RNA of interest are allowed to hybridize onto the membrane, visualized with radiography, and quantified by densitometry. By comparing intensities of bands, one can infer whether a particular gene of interest is being transcribed, and at what relative abundance when compared with other conditions. As the hybridized DNA/RNA complexes are separated by size, Northern blotting can also provide information on the number of isoforms of a particular messenger RNA (mRNA) present in the sample. Within a year of its publication, the protocol was cited 11 times, and 76 more times in the second year, highlighting its popularity and ease of use.

Table 1.1. A comparison of common tools to assess transcriptomics.

	Northern Blot	RT-qPCR	EST	Microarray	RNA-seq
Cost of experiment	Low[a]	Low[b]	High[c]	\$300–700[d]	\$400–1650[e]
Number of genes characterized	1–100 (size of gel comb)[a]	96–384[b]	1000–1500[c]	Full genomes[f]	Full genomes[f]
De novo gene identification	No[a]	No[b]	Yes[c]	Limited, with tiling array[g]	Yes[f]
Equipment barrier	Low[a]	Low[b]	Medium[c]	Medium–high[h]	High (low if sequencing outsourced)[i]
Sample RNA/DNA input	>10 μg[a]	<5 μg[b]	1–5 μg[j]	0.05–5 μg[k]	0.1–1 μg[i]
Dynamic range	Not quantitative[f]	10^5–10^7 [l]	Not quantitative[f]	10^3–10^4 [f]	>10^5 [f]

[a]Su and Huang, 2015
[b]Green and Sambrook, 2018
[c]Whitton *et al.*, 2004
[d]https://www.thermofisher.com/search/browse/results?customGroup=Microorganism+%26+Insect+Expression+Profiling+Arrays+%26+Assays
[e]https://www.abmgood.com/RNA-Sequencing-Service.html
[f]Lowe *et al.*, 2017
[g]Lemetre and Zhang, 2013
[h]https://www.terrauniversal.com/applications/microarray-instruments-supplies-x.php costs of microarrays
[i]https://support.illumina.com/content/dam/illumina-support/documents/documentation/chemistry_documentation/samplepreps_truseq/truseq-stranded-mrna-workflow/truseq-stranded-mrna-workflow-reference-1000000040498-00.pdf
[j]https://www.thermofisher.com/us/en/home/references/protocols/cloning/cloning-protocol/cdna-library-construction.html
[k]Diaz and Barisone, 2011
[l]Morrison *et al.*, 1998

One major limitation of the Northern blot is that it only reveals whether or not a probe is able to hybridize to an RNA target, and provides no information on the strength of the hybridization, which will vary with presence of sequence polymorphism. How would one distinguish whether a probe had complete complementarity to the target, or whether it had perhaps bound to a similar target, as in the case of point mutations? Friedberg *et al.* (1990) asked this very question when they sought to determine whether there was an unknown, previously uncharacterized member of the P450 gene family in rats, and employed a modified version of an endonuclease protection assay for RNA (Myers *et al.*, 1985; Friedberg *et al.*, 1990). In this method, labeled RNA probes antisensed to an mRNA target are created using SP6 or T7 RNA transcription. These probes are added to extracted total mRNA of the sample and allowed to hybridize. RNAse is added at low enough concentrations to only degrade single-stranded RNA (ssRNA), but not protected double-stranded RNA (dsRNA). In addition, RNAse will degrade sites of mismatches in dsRNA; therefore, any probe bound to an RNA transcript that is similar but not identical in sequence will be degraded into reproducible smaller fragments. These fragments can be visualized by gel electrophoresis and the pattern of banding will be specific to each probe-target pair. Using this method, Friedberg *et al.* (1990) were able to identify a new member of the P450II gene family, which had been previously undetected with standard Northern assays using short oligonucleotide probes. In addition to this use, endonuclease protection assays can be used to quantify multiple different targets in one tube. Because band size on the probe is dependent on the length of the probe, by using a combination of probes of different length, it is possible to assay expression levels of multiple genes simultaneously (Stalder *et al.*, 1999).

Both of the previous methods rely on quantification of signal intensity from a radioactively labeled membrane. While quite effective, it has the disadvantage of requiring radioactive handling, and is often not sensitive enough for detecting RNA at extremely low copy number (Fehr *et al.*, 2000; Dean *et al.*, 2002; VanGuilder *et al.*, 2008). The invention of the polymerase chain reaction (PCR) in 1983, coupled with the addition of a heat-stable polymerase in 1988, displayed incredible ability in amplifying DNA sequences, allowing detection of a single DNA copy in a mixture of 10^6 cells (Saiki *et al.*, 1988). The use of reverse transcriptase (RT-PCR) allowed researchers to amplify complementary DNA (cDNA) made from mRNA samples, allowing detection of different RNA isoforms (Mocharla *et al.*, 1990). However, methods to estimate starting concentrations by visualizing final PCR products or competitive PCR were hampered by low sensitivity and accuracy (McCarrey *et al.*, 1992; Piatak *et al.*, 1993). In 1993, Higuchi *et al.* developed a quantitative RT-PCR reaction (qRT-PCR) using fluorescent dyes and a charge-coupled device (CCD) imager. By adding ethidium bromide into the PCR reaction, they could selectively label dsDNA using UV light. As each cycle of PCR approximately doubles the amount of target DNA present, recording the fluorescence after each cycle with an imager allows for accurate quantification of starting template concentration. The highly sensitive nature of PCR means that extremely low mRNA expression can still be measured and quantified. With careful primer design, one can selectively amplify and quantitatively compare different isoforms as well, with a detectable range as low as 16 molecules per reaction (Kubista *et al.*, 2017).

1.2.2 Early transcriptomic methods

So far, the above methods for detecting RNA have been decidedly low-throughput. However, in the 1990s sequencing technology had advanced sufficiently to enable the beginning of transcriptome profiling, making use of expressed sequence tags (ESTs). With this method, extracted mRNA transcripts from experimental samples are converted into bacterial cDNA libraries. A random selection of colonies (around 1000) are selected and sequenced from the 3′ end, generating an EST for each transformant sequenced. These sequences represent expressed genes in the sample and can be used to identify novel genes and coding regions. However, as the relative abundance of an mRNA transcript is proportionally reflected in the number of colonies carrying that EST, EST databases can also be used to provide transcript abundance data of the sample (Audic and Claverie, 1997). EST transcriptome profiling was first used with human liver cells (Okubo *et al.*, 1992), but has been applied to entomology research, such as profiling the transcriptomes of *Toxoptera citricida* (brown citrus aphids) (Hunter *et al.*, 2003) and *Bemisia tabaci* (whitefly) (Leshkowitz *et al.*, 2006). Upon the development of microarray technologies, EST libraries were used to identify novel mRNA sequences to which hybridization probes could be designed (Ote *et al.*, 2004; Guerrero *et al.*, 2009; Bass *et al.*, 2012; Husseneder *et al.*, 2012).

ESTs enabled discovery of novel transcripts, but were resource-intensive due to the costs and labor of sequencing large numbers of bacterial colonies. In order to reduce sequencing load, Serial Analysis of Gene Expression (SAGE) was developed to improve on EST technology (Velculescu *et al.*, 1995). Instead of sequencing 600–800bp long cDNA clones, SAGE used a combination of restriction enzyme digests and ligations to capture 9–14bp 3′ ends of mRNA and clone them into long serial chains. Thus, by sequencing one cDNA clone, one can obtain quantitative data on dozens of transcripts simultaneously, greatly improving throughput as long as the 9–14bp reads are uniquely identifying for each transcript. This is feasible as long as a sequenced genome of the organism or a previously created EST library is available to map the short tags to. A modified version of SAGE, Cap Analysis Gene Expression (CAGE), utilized a similar procedure but captured 5′ ends of RNA transcripts, allowing for discovery of alternative transcription start sites and promoter region identification (Shiraki *et al.*, 2003).

While these sequencing methods jump-started widespread discovery of transcripts, their significant economic and labor costs left room for innovation, particularly when attempting to perform comparative transcriptomics between multiple samples and conditions. In 1995, researchers at Stanford University invented the first microarray, in an effort to allow fast, highly parallelized quantification and comparisons of transcript quantity between samples (Schena *et al.*, 1995). In short, this method utilizes cDNA that is generated from EST or cDNA libraries that are spotted in a known configuration on a glass slide. Next, fluorescent cDNA is generated from extracted mRNA of the samples to be compared. By hybridizing these fluorescent probes to the spotted array and scanning the array with a laser, one is able to determine relative expression levels of the cDNAs in question. By using different fluorescent markers for each sample, multiplexing of both samples onto the same array is possible, and a direct comparison of expression patterns for all genes spotted can be estimated. While the initial technology contained 45 cDNA assays on a single 3.5 × 5.5 mm glass sheet, rapid advances have led to commercially available RNA expression arrays containing over 1 million probes per chip (http://www.affymetrix.com/products_services/arrays/specific/cexpress.affx). One major downside of using expression microarrays is the requirement that the probe targets be known. Typically, a cDNA library is used to create the probes; however, transcripts missing in the library will never be detected. To address this issue, it is possible to use a tiling microarray instead. Rather than specifically targeting known transcripts, tiling microarrays use probes whose collective sequences span large regions of the genome, or even the entire genome sequence (Lemetre and Zhang, 2013). Thus, when extracted mRNA is hybridized, it is possible to detect previously unidentified transcribed regions by mapping the probe sequence back to the genome of the organism. Today, expression microarrays and tiling microarrays can be routinely ordered from and custom generated by a variety of commercial sources, enabling diagnostic as well as research applications, ushering in a modern era of transcriptomics.

1.2.3 Post-genomic

Despite the dramatic increase in parallel detection power and cost reduction, microarrays still left room for improvement. Detecting subtle differences in expression levels using hybridization-based approaches is often challenging, and transcripts expressed at very low levels often go undetected. The advent of cheap, highly parallel, high-throughput sequencing in the late 2000s opened up a new avenue of transcriptomic studies and led to the development of RNA sequencing (RNA-seq), which enables more sensitive detection of rare transcripts and accurate quantification of differential expression (Zhao *et al.*, 2014). The method was first employed in 2006 using 454 sequencing technology (Bainbridge *et al.*, 2006), but the terminology used today was coined in 2008, when several papers published research using the term RNA-seq within several months of each other (Lister *et al.*, 2008; Mortazavi *et al.*, 2008; Nagalakshmi *et al.*, 2008). In its most general format, this method involves generating cDNA from extracted mRNA, sequencing the cDNA, and using alignment software to map reads back to a reference genome that has been previously assembled. If a reference genome is not available, as is the case when working with non-model systems, it is also possible to generate a *de novo* transcriptome assembly by assembling overlapping sequencing reads into complete transcripts, although sequencing errors and repetitive regions or splice variants can hamper this process (Wang and Gribskov, 2017; Geniza and Jaiswal, 2017). After sequencing and mapping has been completed, this data can be used for a variety of purposes, such as quantifying the number of transcripts per gene, differential gene expression between samples, identifying new splice variants or transcription start sites, phylogenetic analysis, and more. Compared with microarrays, RNA-seq offers more flexibility in research applications, as it eliminates the need to design and manufacture custom-printed microarrays.

1.2.4 Quantification of nascent RNA transcription

So far, the techniques covered have focused on capturing steady-state mRNA levels, that is, the current amount of an mRNA transcript in a cell at a particular point in time. What if one desired to understand the rate at which a particular transcript is produced or degraded? While steady-state levels may appear stable, the turnover rate can range widely, with single mRNA molecules exhibiting half-lives ranging from minutes to several hours (Opyrchal *et al.*, 2005). Such information can provide detailed information on transcriptional regulation and kinetics. A common attempt to capture this information across the genome uses chromatin immunoprecipitation of RNA polymerase II (RNAPII) cross-linked to DNA. By using microarrays or sequencing DNA attached to RNAPII and counting the number of reads at a locus, it becomes possible to see a snapshot of the amount of RNAPII presumably actively transcribing RNA (Kim *et al.*, 2005; Muse *et al.*, 2007; Welboren *et al.*, 2009). This method provides a global view of transcription, but has a few limits, namely in its inability to determine which strand is being transcribed, as well as having a resolution of around 100 bp (Schmid *et al.*, 2018). To improve resolution and identify the transcribed strand, several competing techniques arose that centered around chemically labeling nascently transcribed RNA and sequencing them. GRO-seq, developed in 2008, used a nuclear run-on assay, in which nuclei were extracted from flash-frozen samples (Core *et al.*, 2008). *In vitro* transcription is allowed to resume in the presence of a labeled nucleotide (such as 5-bromouridine 5′ triphosphate or biotin-triphosphates) and an inhibitor of transcription initiation. Thus, only genes in the process of being transcribed are labeled, and RNA-seq or microarray data can be used to identify activity. This process allowed strand identification, and a subsequent improvement called PRO-seq used 3′ end sequencing strand-terminating labeled triphosphates to identify RNAP localization on the transcript with base-pair (bp) level resolution (Mahat *et al.*, 2016). An alternative technique, NET-seq, used immunoprecipitation of RNAP followed by 3′ sequencing to accomplish similar results *in vivo* (Churchman and Weissman, 2011).

It is also possible to estimate both the transcription rate and decay rate of an mRNA concurrently in a single experiment. This can be performed by allowing incorporation of labeled nucleotides into new transcripts for a defined amount of time, and quantifying counts of labeled transcripts separately from total RNA counts. Given the ratio of labeled RNA to total RNA in a specified amount of time, it is possible to estimate the mRNA decay rate for each gene, as well as the transcription rate (Dölken *et al.*, 2008; Schmid *et al.*, 2018).

1.2.5 Current state of sequencing technologies for RNA-seq

Currently, the most common technology used to perform sequencing of transcriptomic libraries is provided by the company Illumina, in a method known as sequencing by synthesis, for several reasons. Illumina technology provides highly accurate reads with an average error rate of 0.1% (Ardui *et al.*, 2018), which when combined with its high sequencing output (up to 20 billion reads on the NovaSeq 6000) is critical for correctly determining which gene the transcript sequence maps back to. As long as a high-quality reference genome is available for mapping of sequencing reads, typically 70–90% of reads returned from an Illumina sequencing run are assigned to a gene of origin (Dobin *et al.*, 2013). In addition, the cost per bp for Illumina sequencing is significantly lower than alternative methods, making it desirable for experiments with multiple comparisons and biological replicates.

The major shortcoming of Illumina sequencing technology is the short read length. Illumina offers read lengths up to 300 bp, and thus requires fragmentation of RNA transcripts into short sequences of 150–300 bp. When compared with the average mRNA transcript length in *Drosophila melanogaster* of 3000 bp, this is quite small (Adams *et al.*, 2000). Using short fragments also requires the use of bioinformatic software to 'reconstruct' the mRNA sequences originally present in the sample. This can be achieved by simply counting the number of times a gene gets a read mapped to it (Anders *et al.*, 2015). This method can introduce biases in quantification, since during the fragmentation step longer transcripts will produce more reads mapping to itself than short transcripts. Statistical methods to estimate the original number of molecules of each transcript isoform can be used to account for this, but require knowledge of transcript length, which requires a sequenced genome (Trapnell *et al.*, 2012). This bioinformatic step requires expertise in statistics and computational software, and can significantly increase complexity in analyzing RNA-seq results. Furthermore, not all reads are mapped to unique locations on the genome. When a read maps equally well to multiple genes, there will be ambiguity as to the identity of the transcript. Multi-mapped reads can be discarded during data processing, or assigned in an unbiased manner using modeling approaches (Liao *et al.*, 2014; Anders *et al.*, 2015; Schmid and Grossniklaus, 2015), but either way information may be lost. A survey of best practices for RNA-seq data analysis has been published (Conesa *et al.*, 2016) and the data analysis should be tailored to the research question.

If an experiment only calls for quantification of transcripts present, a cheaper and simpler alternative to full-scale RNA sequencing is 3′ RNA-seq (Torres *et al.*, 2008). Borrowing concepts from SAGE, only the 3′ end of a transcript is sequenced. This removes over-counting bias of long transcripts, as each transcript results in a single read, greatly simplifying bioinformatic analysis. 3′ RNA sequencing is especially useful in measuring transcript abundance in non-model organisms, as the need for a sequenced genome is eliminated (Tandonnet and Torres, 2017). In addition, fewer reads need to be generated, reducing sequencing costs. The major downside of 3′ RNA-seq is that isoform data will not be captured unless the differences in the isoforms are close to the 3′ end. Furthermore, in the case of organisms lacking a sequenced genome, most of the transcript sequence is left unknown.

Several developing technologies today are attempting to circumvent the issues of short reads by applying long-read sequencing technologies to transcriptomics. Pacific Biosciences (PacBio) has recently begun advertising its Single Molecule Real-Time (SMRT) sequencing technology as an effective alternative for RNA-seq experiments, termed Iso-Seq, which was first demonstrated to effectively capture a complete transcriptome by Wang *et al.* (2016). With average read lengths of 30,000 bases at time of writing, entire mRNA molecules can be sequenced without need for fragmentation. While the technology has a higher cost per base pair of sequencing, it essentially eliminates the need for bioinformatic processing to produce analyzable data. In addition, this has strong applications in identifying novel splice variants that may be missed with short-read technology, which requires estimating isoform presence using exon–intron boundaries and other subtle signals (Feng *et al.*, 2011). A competing technology from Oxford Nanopore Technologies also promises incredibly long reads from single molecules, although at present higher error rate and consistency have been a concern (Garalde *et al.*, 2018; Weirather *et al.*, 2017).

1.3 Case Studies of Transcriptomics in Entomology

1.3.1 Development and morphology

When studying development in any organism, transcriptomics provides significant insights where genomics alone might falter. Since the genome of an individual is essentially identical in every cell throughout its lifetime, simply sequencing its genome would provide little context as to what guides the complex stages of development. This is especially relevant when studying holometabolous insects that undergo metamorphosis. Several transcriptomic studies have been dedicated to elucidating these processes by taking a transcriptomic ‘snapshot’ at multiple stages of development (Akbari *et al.*, 2013; Jones *et al.*, 2015; Shrestha *et al.*, 2017). Studies have also been conducted on the earliest stages of embryogenesis to examine the maternal to zygotic transition of RNA, a critical moment in which a developing organism switches from relying on maternally deposited mRNA to transcribing their own endogenous mRNA (Ewen-Campen *et al.*, 2011; Morrow *et al.*, 2014; Ylla *et al.*, 2018). Comparative analysis can even be performed on different organs: Wang *et al.* (2016) sequenced mRNAs from the spinnerets and Filippi's glands of the silkworm, which are both known to be involved in silk production, in an effort to better characterize the contribution of each gland to silk production. Their analysis suggested that Filippi's glands may be supporting silk formation in the spinnerets through the transport of necessary ions and amino acids (Wang *et al.*, 2016).

The field of forensic entomology has made use of unique transcriptomic signatures at each life stage of the blowfly *Calliphora vicina* for determining the minimum time of death (Zajac *et al.*, 2018). Blowflies lay their eggs in dead animals and humans, and have a temperature-regulated life cycle consisting of egg-laying within 24 h, followed by a larval stage in which the larva feeds on the body and steadily grows, before leaving the body and pupating nearby to emerge as an adult blowfly. Traditionally, forensic scientists relied on the length of the larva to estimate its age (Donovan *et al.*, 2006). However, using life-stage-specific gene expression markers to precisely identify age of larva has shown some promise (Tarone and Foran, 2011; Boehme *et al.*, 2013; Abdel-Ghaffer *et al.*, 2014). Zajac *et al.* (2018) generated transcriptomes of individual blowfly larvae at three different incubation temperatures each day from hatching to pupation. With this, they were able to identify genes expressed only at specific developmental windows at extremely high levels. Furthermore, they ensured that expression levels of these markers are independent of the rearing temperatures. This allowed them to develop a qPCR assay of 15 markers that allowed accurate prediction of blowfly larva age to within a day.

1.3.2 Studying phenotype plasticity

Transcriptomics can often reveal much about the adaptability and wide-ranging phenotypes present within a single species that genomics alone would not be able to capture. One of the most striking forms of phenotypic plasticity is diapause, in which insects enter a hibernation-like state of low metabolism and reproduction and enhanced cold tolerance to survive through winter months (Hahn and Denlinger, 2011). Researchers have used transcriptomics to study this phenomenon in a wide range of arthropods, such as moths, mosquitoes, *Drosophila* spp., and mites (Kang *et al.*, 2016; Shearer *et al.*, 2016; Zhao *et al.*, 2017; Deng *et al.*, 2018).

Polyphenism is a phenomenon in which distinct morphs can arise from a genetically similar background. This includes insects that develop highly complex caste systems and live in large social groups. Honey bees have been studied with regard to the development of reproductive systems in worker bees versus queens (Hartfelder *et al.*, 2018), and the key genes have been identified in the development of soldier castes of subterranean termites (Wu *et al.*, 2018). Another excellent example of polyphenism is the swarming locust, *Locusta migratoria*. In this case, locusts can exist as a solitary morph or as a swarming gregarious morph, the latter of which is often a serious concern for agriculture as large swarms devour crops (Roberts, 2015). Seeking to understand the physiological underpinnings of this switch in morphs, the locust genome was recently sequenced and used in conjunction with transcriptome data to identify key differentially regulated genes (Wang *et al.*, 2014). By comparing the transcriptomes of locust nymph brains, nearly 4900 genes were found to be differentially expressed and 45 genes exhibiting alternative splicing. Gregarious morphs expressed more transcripts related to metabolic pathways and synaptic transmission, suggesting high rates of neuronal plasticity.

A transcriptomics approach was also used to assess wing polymorphism in *Gryllus firmus*, the sand field cricket. Sand field crickets exhibit three morphs: a long-winged flying morph, a short-winged flightless morph, and a long-winged morph that loses its ability to fly as an adult through hystolized wing muscles (Zera *et al.*, 1997). The morphs have differential reproductive success, with the flying morph having significantly lower rates of oviposition relative to the non-flying morphs (Zera, 2003, 2009). While many studies have been focused on determining the role of fitness in this phenomenon, fewer efforts have been dedicated to understanding the mechanism underlying this polyphenism (Harrison, 1980; Roff, 1984; Zera *et al.*, 1997). Nanoth Vellichirammal *et al.* (2014) took an RNA-seq approach to this problem to compare the transcriptomes of the morphs. As a reference genome is not yet available for the cricket, a *de novo* transcriptome assembly was generated by taking a subset of the collected samples to create a reference transcriptome for read mapping to enable transcript quantification. To further increase the tissue-specific resolution of their analysis, RNA samples were extracted from dissected wing muscles and fat bodies, focusing on the organs involved in flight and reproduction, respectively. In comparing the two flightless morphs, 49% of the discovered genes were found to be significantly differentially expressed in wing muscles, with most of those being involved in proteolytic pathways up-regulated in the flightless wing muscles. They were also able to detect reduced mitochondrial genes in the flightless morphs and identify genes potentially involved in the increased muscle respiration rates of long-wing flying morphs that had been previously characterized (Zera *et al.*, 1997). In examining the fat bodies, while a much smaller count of differentially expressed genes were identified between long- and short-wing morphs, increased expression of lipid synthesis pathways were detected in long-wing morphs, which coincides with the increased energy storage requirements of flight (Zera and Larsen, 2001). Overall, they identified specific genes that corroborated previous physiological studies, highlighting the power of transcriptomics.

1.3.3 Host switching plasticity

Another area in which transcriptomics excels at providing information is in studying host switching, which is often of interest when investigating invasive species or host–prey interactions. In contrast to the typical 'gene mutation first' theory of adaptation, a more recent alternate theory proposes a 'plasticity first' approach, in which generalist species first use their existing gene pool to occupy a new niche, before eventually evolving specialized genes (De Koning *et al.*, 1980). This suggests a special importance of generalists in the evolution and diversification of organisms. An example of a generalist is *Trichoplusia ni*, the cabbage looper, a moth whose larval stage is a common pest of brassicas such as cabbage, broccoli, and kale. While the looper prefers brassicas, it is able to survive on over 60 species of plants, including celery, tobacco, peas, maize, and tomato (Eichlin and Cunningham, 1978). As many plant species contain their own unique defense mechanisms against insect herbivores, saliva of many herbaceous insects contains compounds to combat plant immune systems, and plants often simultaneously recognize specific insect species by unique proteins in their saliva (De Koning *et al.*, 1980; Maffei *et al.*, 2004; Maischak *et al.*, 2007;Vadassery *et al.*, 2012). Rivera-Vega *et al.* (2017) were interested in determining whether cabbage loopers could change their salivary outputs based on the host they were currently feeding on. This could enable efficient expression of specific defense transcripts unique to the current host, instead of expressing all defensive salivary genes constantly. To this end, Rivera-Vega *et al.* (2017) conducted an RNA-seq analysis on salivary glands of 5th-instar cabbage looper, raised on either live cabbage (preferred host), live tomato (non-preferred), or artificial pinto bean diet (preferred). Since a cabbage looper genome had not been published at the time, reads were grouped *de novo* into gene clusters, and the identified gene sequences were identified by using BLAST (Basic Local Alignment Search Tool) to identify homologs in the *Bombyx mori* (silkworm) genome. From this they identified that while only 4% of total transcripts were differentially regulated between loopers reared on either artificial or cabbage diet, 18% of the transcriptome was differentially regulated between loopers reared on either cabbage or tomato, proving that the salivary gland transcriptome is highly plastic. These included transcripts related to immunity, digestion, and detoxification. While this particular study was only able to identify about half of the genes discovered based on silkworm gene homology, a recently published high-quality genome of cabbage looper would likely further improve identification of differentially expressed genes (Blissard *et al.*, 2018), a perfect example of how differing '-omics' technologies can complement one another.

In contrast to the generalist plasticity observed in the cabbage looper, the parasite *Varroa jacobsoni* (varroa mite) is an example of a specialist that suddenly 'hopped' from parasitizing *Apis cerana* (Asian honeybee) to *Apis mellifera* (European honeybee) as recently as 2008 (Roberts *et al.*, 2015). Previously, *V. jacobsoni* had been observed living on *A. mellifera* brood but being unable to reproduce. The ability of some haplotypes of *V. jacobsoni* to now reproduce on *A. mellifera* alarmingly mimics host switching of *Varroa destructor* from *A. cerana* to *A. mellifera* in the 1950s, an adaptation that has spelled disaster for beekeepers worldwide due to the devastating impact of *V. destructor* on honey bee health today (Ellis and Nalen, 2010; Oldroyd, 1999). Andino *et al.* (2016) attempted to uncover the mechanism by which *V. jacobsoni* transitioned to the European honey bee using RNA-seq. They obtained field RNA samples of *V. jacobsoni* that were reproducing on either *A. cerana* or *A. mellifera*. As a quality draft genome of *V. destructor* was available, Andino *et al.* (2016) were able to combine a *de novo* transcriptome assembly with a guided assembly of *V. jacobsoni* using the *V. destructor* genome, after which reads were mapped back to quantify transcript abundance. By comparing the transcript counts between the two conditions, they conservatively identified 267 differentially expressed genes, with 264 of these being up-regulated in mites reproducing on *A. mellifera*. Of these, increased expression of mitochondrial genes and metabolic pathways were observed, which suggested increased stress on *V. jacobsoni* as they adapt to this new host.

1.3.4 Interspecies comparisons

While differential expression analysis is most commonly applied to a single species being subjected to specific experimental conditions, transcriptomics can also be used to compare the experimental responses of closely related species to understand underlying mechanisms for phenotypic differences. As long as the species are closely related enough to identify homologous genes, differential analysis can be performed by comparing the experimental treatment RNA-seq profile of each species back to a control of the same species. This allows identification of differential expression that is due to treatment conditions, not the inherent differences that might exist between the species (Luo *et al.*, 2014; LoVerso and Cui, 2015; Archer *et al.*, 2016). Such was the approach used by Hoedjes *et al.* (2015) in order to identify key genes involved in long-term memory formation in two species of *Nasonia*, a genus of wasps that parasitize fly larva (Beukeboom and Desplan, 2003). *Nasonia* are able to form short- (STM) and long-term memories (LTM) associating odors with food sources; however, requirements for long-term memory formation differ between species. *Nasonia giraulti* requires multiple conditioning events, spaced several days apart, before a particular odor is associated with prey, while *Nasonia vitripennis* is able to form LTM of an odor with only a single conditioning event (Hoedjes and Smid, 2014). To identify key causative genes for these contrasting phenotypes, Hoedjes *et al.* (2015) exposed both species to a food source with or without a conditioning odor, and after allowing feeding, collected RNA samples from heads at 4 h and 24 h post feeding and performed standard RNA-seq using strand-specific sequencing. Protein-coding transcripts were identified by alignment with a proteome database of *V. vitripennis*, which allowed for determination of sense versus antisense transcripts, while any transcripts not mapping to a protein were checked against a non-coding RNA database. From this analysis, they identified genes that were only differentially expressed in *V. vitripennis* when presented with prey with or without a conditioning odor. These genes, involved in Ras and Rho signaling cascades, glutamate receptors, and PI3K signaling, have previously been associated with LTM in mice and chickadees, suggesting a well-conserved mechanism of LTM. In addition, they discovered genes that were only differentially expressed in *V. giraulti*, which could suggest an LTM inhibitory mechanism.

1.3.5 Phylogenomics

In addition to comparative analysis, transcriptomics has recently been used in the field of phylogenomics (Misof *et al.*, 2014; Wang *et al.*, 2017; Darwell *et al.*, 2018; Fernández *et al.*, 2018; Miller *et al.*, 2018). Up until the mid-20th century, phylogenies were essentially constructed using morphology to distinguish species and determine ancestry history. The advent of sequencing technology in the 1980s enabled researchers to distinguish species and build phylogenetic trees based on DNA 'barcoding' genes that were commonly variable between species, such as the mitochondrial cytochrome oxidase subunit 1 (CO1) gene in animals (Hebert *et al.*, 2004; Smith *et al.*, 2007). By aligning genes between sequences and identifying base substitutions, deletions, and insertions, it became possible to treat each of these changes as independent 'characters'. These molecular characters are unambiguous (A, G, C, or T) and easily converted into data formats amenable to statistical analysis (Brown, 2002). In addition, a single experiment can provide hundreds to hundreds of thousands of traits with ease, depending on the sequencing strategy used. In the past few years, it has become more and more feasible to do large sequencing comparisons of entire genomes or transcriptomes across species to generate more information to build accurate phylogenies. Early large-scale experiments used EST datasets, while more recent experiments use RNA-seq data (Cannon and Kocot, 2016).

Transcriptomic methods have been applied in identifying phylogenetic relationships between all orders of insects. The earliest records of insect fossils date to approximately 400 million years ago in the Devonian period, with no fossils existing before this. However, some of the fossil data, as well as previous molecular phylogenetics work, suggested that insects arose earlier, between the Silurian or even Cambrian period (Engel and Grimaldi, 2004; Rota-Stabelli *et al.*, 2013). In an attempt to resolve these discrepancies, 102 insect transcriptomes were sequenced *de novo*, spanning all known orders of insects (Misof *et al.*, 2014). From this massive amount of data, they used 1478 genes to generate a phylogenetic tree of all insects, using both nucleotide sequence-based methods as well as amino acid sequence-based method. Their results indicated that insects originated in the early Ordovician period, essentially concurrently with or just after land colonization by plants (Morris *et al.*, 2018). They also provided evidence supporting previous theories on the evolution of flight during the Devonian period (Engel and Grimaldi, 2004), and helped to corroborate previous phylogenetic groupings of silverfish, Holometabola, and Polyneoptera (Wiegmann *et al.*, 2009; Ishiwata *et al.*, 2011; Yoshizawa, 2011; Blanke *et al.*, 2014).

Once these massive sets of data are published, they can easily be leveraged by other researchers for a variety of further questions, such as tracing the evolution of a specific gene. One such gene of interest is *doublesex* (*dsx*), which has been identified as responsible for sex determination in several orders of insects as well as crustaceans and arachnids (Coschigano and Wensink, 1993; Toyota *et al*., 2013; Wang *et al*., 2014; Pomerantz and Hoy, 2015; Ledón-Rettig *et al*., 2017). In insects, alternative splicing of *dsx* leads to sex-specific development, while in the crustacean *Daphnia magna* and arachnid *Metaseiulus occidentalis* this is not the case. Price *et al*. (2015) wanted to determine how prevalent *dsx* was across all orders of insects by accessing publicly available transcriptome data, including the Misof *et al*. (2014) data mentioned above. From their analysis, it was revealed that *dsx* was present in nearly every order of insect. In addition, evidence of alternative splicing was found in several orders, including basal lineages, suggesting alternative splicing as a mechanism for sex determination was present in the most recent common ancestor of insects. They were also able to detect a loss in gene copy; both *D. magna* and *M. occidentalis* contain two copies of *dsx*, while only seven of the 32 insect orders had evidence of a second gene copy.

1.4 Conclusion

While powerful strides are being made today in elucidating insect morphology, evolution, and development, future technology and innovation promise much greater advances in knowledge. Long molecule sequencing, now and in the near future, will allow for improved read mapping and identification of isoforms. These new technologies will also reduce steps in RNA-seq bioinformatic analysis, a current major bottleneck in many of these studies. The advent of single-cell RNA sequencing, such as the 10X Genomics Chromium technology, will continue to improve resolution of RNA-seq and provide unprecedented cell-type specific transcriptomic profiling without the need for intensive manual dissections or preparative work. With time, these techniques will result in less effort being spent on sample preparation and bioinformatic preprocessing, and more effort dedicated to better annotation and elucidation of gene function.

Acknowledgements

Research in our laboratory is supported by NIH R01 GM102225, NSF IOS 1456297, USDA SCRI 2015-51181-24252, and USDA NIFA CA-D-ENM-2150-H awarded to JCC.

References

Abdel-Ghaffer, H., Abd-Elmottelb, M. and Zaky, A. (2014) Age-dependent gene expression of blow fly *Lucilia cuprina (Diptera: Calliphoridae)* during egg development improving age estimation in forensic entomology. *Journal of Forensic Research* 5(5), 1. DOI: 10.4172/2157-7145.1000238.

Adams, M.D., Celniker, S.E., Holt, R.A., Evans, C.A., Gocayne, J.D. *et al.* (2000) The genome sequence of *Drosophila melanogaster*. *Science* 287(5461), 2185–2195. DOI: 10.1126/science.287.5461.2185.

Akbari, O.S., Antoshechkin, I., Sandler, J., Hay, B.A., Williams, B. *et al.* (2013) The developmental transcriptome of the mosquito Aedes aegypti, an invasive species and major arbovirus vector. Genes | Genomes | . *Genetics* 3, 1493–1509.

Alwine, J.C., Kemp, D.J. and Stark, G.R. (1977) Method for detection of specific RNAs in agarose gels by transfer to diazobenzyloxymethyl-paper and hybridization with DNA probes. *Proceedings of the National Academy of Sciences* 74(12), 5350–5354. DOI: 10.1073/pnas.74.12.5350.

Anders, S., Pyl, P.T. and Huber, W. (2015) HTSeq-A Python framework to work with high-throughput sequencing data. *Bioinformatics* 31(2), 166–169. DOI: 10.1093/bioinformatics/btu638.

Andino, G.K., Gribskov, M., Anderson, D.L., Evans, J.D. and Hunt, G.J. (2016) Differential gene expression in *Varroa jacobsoni* mites following a host shift to European honey bees (*Apis mellifera*). *BMC Genomics* 17(1), 926. DOI: 10.1186/s12864-016-3130-3.

Archer, D.B., Velasco de Castro Oliveira, J., van Munster, J.M., Blythe, M.J., Sang, F. *et al.* (2016) Transcriptomic responses of mixed cultures of ascomycete fungi to lignocellulose using dual RNA-seq reveal inter-species antagonism and limited beneficial effects on CAZyme expression. *Fungal Genetics and Biology* 102, 4–21.

Ardui, S., Ameur, A., Vermeesch, J.R. and Hestand, M.S. (2018) Single molecule real-time (SMRT) sequencing comes of age: applications and utilities for medical diagnostics. *Nucleic Acids Research* 46(5), 2159–2168. DOI: 10.1093/nar/gky066.

Audic, S. and Claverie, J.M. (1997) The significance of digital gene expression profiles. *Genome Research* 7(10), 986–995. DOI: 10.1101/gr.7.10.986.

Bainbridge, M.N., Warren, R.L., Hirst, M., Romanuik, T., Zeng, T. *et al.* (2006) Analysis of the prostate cancer cell line LNCaP transcriptome using a sequencing-by-synthesis approach. *BMC Genomics* 7(1), 246. DOI: 10.1186/1471-2164-7-246.

Bass, C., Hebsgaard, M.B. and Hughes, J. (2012) Genomic resources for the brown planthopper, *Nilaparvata lugens*: transcriptome pyrosequencing and microarray design. *Insect Science* 19(1), 1–12. DOI: 10.1111/j.1744-7917.2011.01440.x.

Beukeboom, L. and Desplan, C. (2003) Nasonia. *Current Biology* 13(22), R860–R860. DOI: 10.1016/j.cub.2003.10.042.

Blanke, A., Koch, M., Wipfler, B., Wilde, F. and Misof, B. (2014) Head morphology of *Tricholepidion gertschi* indicates monophyletic *Zygentoma*. *Frontiers in Zoology* 11(1), 16. DOI: 10.1186/1742-9994-11-16.

Blissard, G., Chen, W., Wang, P., Hegedus, D., Song, X. *et al.* (2018) A high-quality chromosome-level genome assembly of a generalist herbivore, *Trichoplusia ni*. *Molecular Ecology Resources* 19, 485–496.

Boehme, P., Spahn, P., Amendt, J. and Zehner, R. (2013) Differential gene expression during metamorphosis: a promising approach for age estimation of forensically important *Calliphora vicina pupae (Diptera: Calliphoridae)*. *International Journal of Legal Medicine* 127(1), 243–249. DOI: 10.1007/s00414-012-0699-1.

Brown, T. (2002) *Genomes*. Wiley-Liss, Oxford, UK.

Cannon, J.T. and Kocot, K.M. (2016) Phylogenomics using transcriptome data. *Methods in Molecular Biology* 1452, 65–80.

Churchman, L.S. and Weissman, J.S. (2011) Nascent transcript sequencing visualizes transcription at nucleotide resolution. *Nature* 469(7330), 368–373. DOI: 10.1038/nature09652.

Conesa, A., Madrigal, P., Tarazona, S., Gomez-Cabrero, D., Cervera, A. *et al.* (2016) A survey of best practices for RNA-seq data analysis. *Genome Biology* 17(1), 13. DOI: 10.1186/s13059-016-0881-8.

Core, L.J., Waterfall, J.J. and Lis, J.T. (2008) Nascent RNA sequencing reveals widespread pausing and divergent initiation at human promoters. *Science* 322(5909), 1845–1848. DOI: 10.1126/science.1162228.

Coschigano, K.T. and Wensink, P.C. (1993) Sex-specific transcriptional regulation by the male and female doublesex proteins of *Drosophila*. *Genes and Development* 7(1), 42–54. DOI: 10.1101/gad.7.1.42.

Darwell, C.T., Ayyampalayam, S., Leebens-Mack, J., Smith, C.I., Segraves, K.A. *et al.* (2018) Phylogenomic reconstruction of transcriptome data confirms the basal position of *Prodoxidae* moths within the order *Lepidoptera*. *Arthropod Systematics and Phylogeny* 76, 59–64.

De Koning, A.J., Budzelaar, P.H.M., Boersma, J. and van der Kerk, G.J.M. (1980) Specific and selective reduction of aromatic nitrogen heterocycles with the bis-pyridine complexes of bis(1,4-dihydro-1-pyridyl)zinc and bis(1,4-dihydro-1-pyridyl)magnesium. *Journal of Organometallic Chemistry* 199(2), 153–169. DOI: 10.1016/S0022-328X(00)83849-8.

Dean, J.D., Goodwin, P.H. and Hsiang, T. (2002) Comparison of relative RT-PCR and northern blot analyses to measure expression of β-1,3-glucanase in *Nicotiana benthamiana* infected with *Colltotrichum destructivum*. *Plant Molecular Biology Reporter* 20(4), 347–356. DOI: 10.1007/BF02772122.

Deng, Y., Li, F., Rieske, L.K., Sun, L.-L. and Sun, S.-H. (2018) Transcriptome sequencing for identification of diapause-associated genes in fall webworm, *Hyphantria cunea Drury*. *Gene* 668, 229–236. DOI: 10.1016/j.gene.2018.05.023.

Diaz, E. and Barisone, G.A. (2011) DNA Microarrays: Sample quality control, array hybridization and scanning. *Journal of Visualized Experiments* 49(49), 2546. DOI: 10.3791/2546.

Dobin, A., Davis, C.A., Schlesinger, F., Drenkow, J., Zaleski, C. *et al.* (2013) STAR: Ultrafast universal RNA-seq aligner. *Bioinformatics* 29(1), 15–21. DOI: 10.1093/bioinformatics/bts635.

Dölken, L., Ruzsics, Z., Rädle, B., Friedel, C.C., Zimmer, R. *et al.* (2008) High-Resolution gene expression profiling for simultaneous kinetic parameter analysis of RNA synthesis and decay. *RNA* 14(9), 1959–1972. DOI: 10.1261/rna.1136108.

Donovan, S.E., Hall, M.J.R., Turner, B.D. and Moncrieff, C.B. (2006) Larval growth rates of the blowfly, *Calliphora vicina*, over a range of temperatures. *Medical and Veterinary Entomology* 20(1), 106–114. DOI: 10.1111/j.1365-2915.2006.00600.x.

Eichlin, T.D. and Cunningham, H.B. (1978) The Plusiinae (Lepidoptera: Noctuidae) of America north of Mexico, emphasizing genitalic and larval morphology. *Technical Bulletin 1567 14–15*, US Department of Agriculture, Washington, DC.

Ellis, J.D. and Nalen, C.M.Z. (2010) Varroa Mite, *Varroa destructor* Anderson and Trueman. Entomology and Nematology Dept, UF/IFAS Extension. University of Florida, Gainesville, Florida.

Engel, M.S. and Grimaldi, D.A. (2004) New light shed on the oldest insect. *Nature* 427(6975), 627–630. DOI: 10.1038/nature02291.

Ewen-Campen, B., Shaner, N., Panfilio, K.A., Suzuki, Y., Roth, S. *et al.* (2011) The maternal and early embryonic transcriptome of the milkweed bug *Oncopeltus fasciatus*. *BMC Genomics* 12(1), 61. DOI: 10.1186/1471-2164-12-61.

Fehr, J.E., Trotter, G.W., Oxford, J.T. and Hart, D.A. (2000) Comparison of Northern blot hybridization and a reverse transcriptase-polymerase chain reaction technique for measurement of mRNA expression of metalloproteinases and matrix components in articular cartilage and synovial membrane from horses with osteoarthritis. *American Journal of Veterinary Research* 61(8), 900–905. DOI: 10.2460/ajvr.2000.61.900.

Feng, J., Li, W. and Jiang, T. (2011) Inference of isoforms from short sequence reads. *Journal of Computational Biology* 18(3), 305–321. DOI: 10.1089/cmb.2010.0243.

Fernández, R., Kallal, R.J., Dimitrov, D., Ballesteros, J.A., Arnedo, M.A. *et al.* (2018) Phylogenomics, diversification dynamics, and comparative transcriptomics across the spider tree of life. *Current Biology* 28(13), 2190–2193. DOI: 10.1016/j.cub.2018.06.018.

Friedberg, T., Grassow, M.A. and Oesch, F. (1990) Selective detection of mRNA forms encoding the major phenobarbital inducible cytochromes P450 and other members of the P450IIB family by the RNAse A protection assay. *Archives of Biochemistry and Biophysics* 279(1), 167–173. DOI: 10.1016/0003-9861(90)90477-G.

Garalde, D.R., Snell, E.A., Jachimowicz, D., Sipos, B., Lloyd, J.H. *et al.* (2018) Highly parallel direct RNA sequencing on an array of nanopores. *Nature Methods* 15(3), 201–206. DOI: 10.1038/nmeth.4577.

Geniza, M. and Jaiswal, P. (2017) Tools for building de novo transcriptome assembly. *Current Plant Biology* 11–12, 41–45. DOI: 10.1016/j.cpb.2017.12.004.

Green, M.R. and Sambrook, J. (2018) Quantification of RNA by Real-Time Reverse Transcription-Polymerase Chain Reaction (RT-PCR). *Cold Spring Harbor Protocols* 2018(10), pdb.prot095042. DOI: 10.1101/pdb.prot095042.

Guerrero, F.D., Dowd, S.E., Sun, Y., Saldivar, L., Wiley, G.B. *et al.* (2009) Microarray analysis of female- and larval-specific gene expression in the horn fly (Diptera: Muscidae). *Journal of Medical Entomology* 46(2), 257–270. DOI: 10.1603/033.046.0210.

Hahn, D.A. and Denlinger, D.L. (2011) Energetics of insect diapause. *Annual Review of Entomology* 56(1), 103–121. DOI: 10.1146/annurev-ento-112408-085436.

Hargreaves, A.D. and Mulley, J.F. (2015) Assessing the utility of the Oxford Nanopore MinION for snake venom gland cDNA sequencing. *PeerJ* 3(3), e1441. DOI: 10.7717/peerj.1441.

Harrison, R.G. (1980) Dispersal polymorphisms in insects. *Annual Review of Ecology and Systematics* 11(1), 95–118. DOI: 10.1146/annurev.es.11.110180.000523.

Hartfelder, K., Tiberio, G.J., Lago, D.C., Dallacqua, R.P. and Bitondi, M.M.G. (2018) The ovary and its genes – developmental processes underlying the establishment and function of a highly divergent reproductive system in the female castes of the honey bee, *Apis mellifera. Apidologie* 49(1), 49–70. DOI: 10.1007/s13592-017-0548-9.

Hebert, P.D.N., Penton, E.H., Burns, J.M., Janzen, D.H. and Hallwachs, W. (2004) Ten species in one: DNA barcoding reveals cryptic species in the neotropical skipper butterfly *Astraptes fulgerator. Proceedings of the National Academy of Sciences* 101(41), 14812–14817. DOI: 10.1073/pnas.0406166101.

Higuchi, R., Fockler, C., Dollinger, G. and Watson, R. (1993) Kinetic PCR analysis: real-time monitoring of DNA amplification reactions. *Nature Biotechnology* 11(9), 1026–1030. DOI: 10.1038/nbt0993-1026.

Hoedjes, K.M. and Smid, H.M. (2014) Natural variation in long-term memory formation among *Nasonia* parasitic wasp species. *Behavioural Processes* 105, 40–45. DOI: 10.1016/j.beproc.2014.02.014.

Hoedjes, K.M., Smid, H.M., Schijlen, E.G.W.M., Vet, L.E.M. and van Vugt, J.J.F.A. (2015) Learning-induced gene expression in the heads of two *Nasonia* species that differ in long-term memory formation. *BMC Genomics* 16(1), 162. DOI: 10.1186/s12864-015-1355-1.

Hunter, W.B., Dang, P.M., Bausher, M.G., Chaparro, J.X., McKendree, W. *et al.* (2003) Aphid biology: expressed genes from alate *Toxoptera citricida*, the brown citrus aphid. *Journal of Insect Science* 3(23), 1–7. DOI: 10.1673/031.003.2301.

Husseneder, C., McGregor, C., Lang, R.P., Collier, R. and Delatte, J. (2012) Transcriptome profiling of female alates and egg-laying queens of the Formosan subterranean termite. *Comparative Biochemistry and Physiology. Part D, Genomics & Proteomics* 7, 14–27.

i5k Consortium (2013) The i5K initiative: advancing arthropod genomics for knowledge, human health, agriculture, and the environment. *Journal of Heredity* 104(5), 595–600. DOI: 10.1093/jhered/est050.

Ishiwata, K., Sasaki, G., Ogawa, J., Miyata, T. and Su, Z.-H. (2011) Phylogenetic relationships among insect orders based on three nuclear protein-coding gene sequences. *Molecular Phylogenetics and Evolution* 58(2), 169–180. DOI: 10.1016/j.ympev.2010.11.001.

Jones, B.M., Wcislo, W.T. and Robinson, G.E. (2015) Developmental transcriptome for a facultatively eusocial bee, *Megalopta genalis*. *Genes | Genomes | . Genetics* 5, 2127–2135.

Kang, D.S., Cotten, M.A., Denlinger, D.L. and Sim, C. (2016) Comparative transcriptomics reveals key gene expression differences between diapausing and non-diapausing adults of *Culex pipiens*. *PLoS ONE* 11(4), e0154892. DOI: 10.1371/journal.pone.0154892.

Kim, T.H., Barrera, L.O., Zheng, M., Qu, C., Singer, M.A. *et al.* (2005) A high-resolution map of active promoters in the human genome. *Nature* 436(7052), 876–880. DOI: 10.1038/nature03877.

Kubista, M., Sjögreen, B., Björkman, J., Forootan, A., Sjöback, R. *et al.* (2017) Methods to determine limit of detection and limit of quantification in quantitative real-time PCR (qPCR). *Biomolecular Detection and Quantification* 12, 1–6.

Ledón-Rettig, C.C., Zattara, E.E. and Moczek, A.P. (2017) Asymmetric interactions between doublesex and tissue- and sex-specific target genes mediate sexual dimorphism in beetles. *Nature Communications* 8(1), 14593. DOI: 10.1038/ncomms14593.

Lemetre, C. and Zhang, Z.D. (2013) A brief introduction to tiling microarrays: principles, concepts, and applications. In: Lee, T.L. and Shui Luk, A.C. (eds) *Tiling Arrays: Methods and Protocols*. Humana Press, Totowa, New Jersey, pp. 3–19.

Leshkowitz, D., Gazit, S., Reuveni, E., Ghanim, M., Czosnek, H. *et al.* (2006) Whitefly (*Bemisia tabaci*) genome project: analysis of sequenced clones from egg, instar, and adult (viruliferous and non-viruliferous) cDNA libraries. *BMC Genomics* 7(1), 79. DOI: 10.1186/1471-2164-7-79.

Lewin, H.A., Robinson, G.E., Kress, W.J., Baker, W.J., Coddington, J. *et al.* (2018) Earth BioGenome Project: sequencing life for the future of life. *Proceedings of the National Academy of Sciences* 115(17), 4325–4333. DOI: 10.1073/pnas.1720115115.

Liao, Y., Smyth, G.K. and Shi, W. (2014) FeatureCounts: an efficient general purpose program for assigning sequence reads to genomic features. *Bioinformatics* 30(7), 923–930. DOI: 10.1093/bioinformatics/btt656.

Lister, R., O'Malley, R.C., Tonti-Filippini, J., Gregory, B.D., Berry, C.C. *et al.* (2008) Highly integrated single-base resolution maps of the epigenome in Arabidopsis. *Cell* 133(3), 523–536. DOI: 10.1016/j.cell.2008.03.029.

LoVerso, P.R. and Cui, F. (2015) A computational pipeline for cross-species analysis of RNA-seq data using R and Bioconductor. *Bioinformatics and Biology Insights* 9, 165–174. DOI: 10.4137/BBI.S30884.

Lowe, R., Shirley, N., Bleackley, M., Dolan, S. and Shafee, T. (2017) Transcriptomics technologies. *PLoS Computational Biology* 13(5), e1005457. DOI: 10.1371/journal.pcbi.1005457.

Luo, Y., Li, C., Landis, A.G., Wang, G., Stoeckel, J. *et al.* (2014) Transcriptomic profiling of differential responses to drought in two freshwater mussel species, the giant floater *Pyganodon grandis* and the pondhorn *Uniomerus tetralasmus*. *PLoS ONE* 9(2), e89481. DOI: 10.1371/journal.pone.0089481.

Maffei, M., Bossi, S., Spiteller, D., Mithöfer, A. and Boland, W. (2004) Effects of feeding *Spodoptera littoralis* on lima bean leaves. I. Membrane potentials, intracellular calcium variations, oral secretions, and regurgitate components. *Plant Physiology* 134(4), 1752–1762. DOI: 10.1104/pp.103.034165.

Mahat, D.B., Kwak, H., Booth, G.T., Jonkers, I.H., Danko, C.G. *et al.* (2016) Base-pair-resolution genome-wide mapping of active RNA polymerases using precision nuclear run-on (PRO-seq). *Nature Protocols* 11(8), 1455–1476. DOI: 10.1038/nprot.2016.086.

Maischak, H., Grigoriev, P.A., Vogel, H., Boland, W. and Mithöfer, A. (2007) Oral secretions from herbivorous lepidopteran larvae exhibit ion channel-forming activities. *FEBS Letters* 581(5), 898–904. DOI: 10.1016/j.febslet.2007.01.067.

McCarrey, J.R., Dilworth, D.D. and Sharp, R.M. (1992) Semiquantitative analysis of X-linked gene expression during spermatogenesis in the mouse: ethidium-bromide staining of RT-PCR products. *Genetic Analysis, Techniques and Applications* 9(4), 117–123. DOI: 10.1016/1050-3862(92)90051-6.

Miller, D.B., Bartlett, S., Sartori, M., Breinholt, J.W. and Ogden, T.H. (2018) Anchored phylogenomics of burrowing mayflies (*Ephemeroptera*) and the evolution of tusks. *Systematic Entomology* 43(4), 692–701. DOI: 10.1111/syen.12298.

Misof, B., Liu, S., Meusemann, K., Peters, R.S., Donath, A. *et al.* (2014) Phylogenomics resolves the timing and pattern of insect evolution. *Science* 346(6210), 763–767. DOI: 10.1126/science.1257570.

Mocharla, H., Mocharia, R. and Hodes, M. (1990) Coupled reverse transcription-polymerase chain reaction (RT-PCR) as a sensitive and rapid method for isozyme genotyping. *Gene* 93(2), 271–275. DOI: 10.1016/0378-1119(90)90235-J.

Morris, J.L., Puttick, M.N., Clark, J.W., Edwards, D., Kenrick, P. *et al.* (2018) The timescale of early land plant evolution. *Proceedings of the National Academy of Sciences* 115(10), E2274–E2283. DOI: 10.1073/pnas.1719588115.

Morrison, T.B., Weis, J.J. and Wittwer, C.T. (1998) Quantification of lowcopy transcripts by continuous SYBR® green I monitoring during amplification. *BioTechniques* 24, 954–962.

Morrow, J.L., Riegler, M., Gilchrist, A., Shearman, D.C.A. and Frommer, M. (2014) Comprehensive transcriptome analysis of early male and female *Bactrocera jarvisi* embryos. *BMC Genetics* 15(Suppl. 2), S7. DOI: 10.1186/1471-2156-15-S2-S7.

Mortazavi, A., Williams, B.A., McCue, K., Schaeffer, L. and Wold, B. (2008) Mapping and quantifying mammalian transcriptomes by RNA-Seq. *Nature Methods* 5(7), 621–628. DOI: 10.1038/nmeth.1226.

Muse, G.W., Gilchrist, D.A., Nechaev, S., Shah, R., Parker, J.S. *et al.* (2007) RNA polymerase is poised for activation across the genome. *Nature Genetics* 39(12), 1507–1511. DOI: 10.1038/ng.2007.21.

Myers, R.M., Larin, Z. and Maniatis, T. (1985) Detection of single base substitutions by ribonuclease cleavage at mismatches in RNA:DNA duplexes. *Science* 230(4731), 1242–1246. DOI: 10.1126/science.4071043.

Nagalakshmi, U., Wang, Z., Waern, K., Shou, C., Raha, D. *et al.* (2008) The transcriptional landscape of the yeast genome defined by RNA sequencing. *Science* 320(5881), 1344–1349. DOI: 10.1126/science.1158441.

Nanoth Vellichirammal, N., Zera, A.J., Schilder, R.J., Wehrkamp, C., Riethoven, J.-J.M. *et al.* (2014) De novo transcriptome assembly from fat body and flight muscles transcripts to identify morph-specific gene expression profiles in *Gryllus firmus*. *PLoS ONE* 9(1), e82129. DOI: 10.1371/journal.pone.0082129.

Okubo, K., Hori, N., Matoba, R., Niiyama, T., Fukushima, A. *et al.* (1992) Large scale cDNA sequencing for analysis of quantitative and qualitative aspects of gene expression. *Nature Genetics* 2(3), 173–179. DOI: 10.1038/ng1192-173.

Oldroyd, B.P. (1999) Coevolution while you wait: *Varroa jacobsoni*, a new parasite of western honeybees. *Trends in Ecology & Evolution* 14(8), 312–315. DOI: 10.1016/S0169-5347(99)01613-4.

Opyrchal, M., Anderson, J.R., Sokoloski, K.J., Wilusz, C.J. and Wilusz, J. (2005) A cell-free mRNA stability assay reveals conservation of the enzymes and mechanisms of mRNA decay between mosquito and mammalian cell lines. *Insect Biochemistry and Molecular Biology* 35(12), 1321–1334. DOI: 10.1016/j.ibmb.2005.08.004.

Ote, M., Mita, K., Kawasaki, H., Seki, M., Nohata, J. *et al.* (2004) Microarray analysis of gene expression profiles in wing discs of *Bombyx mori* during pupal ecdysis. *Insect Biochemistry and Molecular Biology* 34(8), 775–784. DOI: 10.1016/j.ibmb.2004.04.002.

Piatak, M., Luk, K.C., Williams, B. and Lifson, J.D. (1993) Quantitative competitive polymerase chain reaction for accurate quantitation of HIV DNA and RNA species. *BioTechniques* 14, 70–81.

Pomerantz, A.F. and Hoy, M.A. (2015) Expression analysis of *Drosophila* doublesex, transformer-2, intersex, fruitless-like, and vitellogenin homologs in the parahaploid predator *Metaseiulus occidentalis (Chelicerata: Acari: Phytoseiidae). Experimental and Applied Acarology* 65(1), 1–16. DOI: 10.1007/s10493-014-9855-2.

Price, D.C., Egizi, A. and Fonseca, D.M. (2015) The ubiquity and ancestry of insect doublesex. *Scientific Reports* 5(1), 13068. DOI: 10.1038/srep13068.

Rivera-Vega, L.J., Galbraith, D.A., Grozinger, C.M. and Felton, G.W. (2017) Host plant driven transcriptome plasticity in the salivary glands of the cabbage looper (*Trichoplusia ni*). *PLoS ONE* 12(8), e0182636. DOI: 10.1371/journal.pone.0182636.

Roberts, J.M.K., Anderson, D.L. and Tay, W.T. (2015) Multiple host shifts by the emerging honeybee parasite, *Varroa jacobsoni. Molecular Ecology* 24(10), 2379–2391. DOI: 10.1111/mec.13185.

Roberts, K. (2015) NSW department of primary industries. locust damage in crops and pastures. Available at: https://www.dpi.nsw.gov.au/_data/assets/pdf_file/0005/336983/Locust-damage-in-crops-and-pastures.pdf (accessed 21 June 2019).

Roff, D.A. (1984) The cost of being able to fly: a study of wing polymorphism in two species of crickets. *Oecologia* 63(1), 30–37. DOI: 10.1007/BF00379781.

Rota-Stabelli, O., Daley, A. C. and Pisani, D. (2013) Molecular timetrees reveal a Cambrian colonization of land and a new scenario for ecdysozoan evolution. *Current Biology* 23(5), 392–398. DOI: 10.1016/j.cub.2013.01.026.

Saiki, R., Gelfand, D., Stoffel, S., Scharf, S., Higuchi, R. *et al.* (1988) Primer-directed enzymatic amplification of DNA with a thermostable DNA polymerase. *Science* 239(4839), 487–491. DOI: 10.1126/science.239.4839.487.

Schena, M., Shalon, D., Davis, R. and Brown, P.O.R. (1995) Quantitative monitoring of gene expression patterns with a complementary DNA microarray. *Science* 270(5235), 467–470. DOI: 10.1126/science.270.5235.467.

Schmid, M.W. and Grossniklaus, U. (2015) Rcount: Simple and flexible RNA-Seq read counting. *Bioinformatics* 31(3), 436–437. DOI: 10.1093/bioinformatics/btu680.

Schmid, M., Tudek, A. and Jensen, T.H. (2018) Simultaneous measurement of transcriptional and post-transcriptional parameters by 3′ end RNA-Seq. *Cell Reports* 24(9), 2468–2478. DOI: 10.1016/j.celrep.2018.07.104.

Shearer, P.W., West, J.D., Walton, V.M., Brown, P.H., Svetec, N. *et al.* (2016) Seasonal cues induce phenotypic plasticity of *Drosophila suzukii* to enhance winter survival. *BMC Ecology* 16(1), 11. DOI: 10.1186/s12898-016-0070-3.

Shiraki, T., Kondo, S., Katayama, S., Waki, K., Kasukawa, T. *et al.* (2003) Cap analysis gene expression for high-throughput analysis of transcriptional starting point and identification of promoter usage. *Proceedings of the National Academy of Sciences of the United States of America* 100(26), 15776–15781. DOI: 10.1073/pnas.2136655100.

Shrestha, A., Champagne, D.E., Culbreath, A.K., Rotenberg, D., Whitfield, A.E. *et al.* (2017) Transcriptome changes associated with Tomato spotted wilt virus infection in various life stages of its thrips vector, *Frankliniella fusca (Hinds). Journal of General Virology* 98(8), 2156–2170. DOI: 10.1099/jgv.0.000874.

Smith, M.A., Wood, D.M., Janzen, D.H., Hallwachs, W. and Hebert, P.D.N. (2007) DNA barcodes affirm that 16 species of apparently generalist tropical parasitoid flies (*Diptera, Tachinidae*) are not all generalists. *Proceedings of the National Academy of Sciences* 104(12), 4967–4972. DOI: 10.1073/pnas.0700050104.

Stalder, A.K., Pagenstecher, A., Kincaid, C.L. and Campbell, Iain L. (1999) Analysis of gene expression by multiprobe RNase protection assay. In: Harry, J. and Tilson, H.A. (eds) *Neurodegeneration Methods and Protocols*. Humana Press, Totowa, New Jersey, pp. 53–66.

Su, S. and Huang, D. (2015) Northern blot. *Protocol Exchange*. DOI: 10.1038/protex.2015.080.

Tandonnet, S. and Torres, T.T. (2017) Traditional versus 3′ RNA-seq in a non-model species. *Genomics Data* 11, 9–16. DOI: 10.1016/j.gdata.2016.11.002.

Tarone, A.M. and Foran, D.R. (2011) Gene expression during blow fly development: improving the precision of age estimates in forensic entomology. *Journal of Forensic Sciences* 56(Suppl. 1), S112–122. DOI: 10.1111/j.1556-4029.2010.01632.x.

Thomas, G.W.C., Dohmen, E., Hughes, D.S.T., Murali, S.C., Poelchau, M. *et al.* (2018) The genomic basis of arthropod diversity. *BioRxiv Genomics* 382945.

Torres, T.T., Metta, M., Ottenwälder, B. and Schlötterer, C. (2008) Gene expression profiling by massively parallel sequencing. *Genome Research* 18(1), 172–177. DOI: 10.1101/gr.6984908.

Toyota, K., Kato, Y., Sato, M., Sugiura, N., Miyagawa, S. *et al.* (2013) Molecular cloning of doublesex genes of four *Cladocera* (water flea) species. *BMC Genomics* 14(1), 239. DOI: 10.1186/1471-2164-14-239.

Trapnell, C., Hendrickson, D.G., Sauvageau, M., Goff, L., Rinn, J.L. *et al.* (2012) Differential analysis of gene regulation at transcript resolution with RNA-seq. *Nature Biotechnology* 31(1), 46–53. DOI: 10.1038/nbt.2450.

Vadassery, J., Scholz, S.S. and Mithöfer, A. (2012) Multiple calmodulin-like proteins in *Arabidopsis* are induced by insect-derived (Spodoptera littoralis) oral secretion. *Plant Signaling & Behavior* 7(10), 1277–1280. DOI: 10.4161/psb.21664.

VanGuilder, H.D., Vrana, K.E. and Freeman, W.M. (2008) Twenty-five years of quantitative PCR for gene expression analysis. *BioTechniques* 44(5), 619–626. DOI: 10.2144/000112776.

Velculescu, V.E., Zhang, L., Vogelstein, B. and Kinzler, K.W. (1995) Serial analysis of gene expression. *Science* 270(5235), 484–487. DOI: 10.1126/science.270.5235.484.

Wang, X., Li, Y., Peng, L., Chen, H., Xia, Q. *et al.* (2016) Comparative transcriptome analysis of *Bombyx mori* spinnerets and Filippi's glands suggests their role in silk fiber formation. *Insect Biochemistry and Molecular Biology* 68, 89–99. DOI: 10.1016/j.ibmb.2015.11.003.

Wang, X.-Y., Zheng, Z.-Z., Song, H.-S. and Xu, Y.-Z. (2014) Conserved RNA cis-elements regulate alternative splicing of *Lepidopteran* doublesex. *Insect Biochemistry and Molecular Biology* 44, 1–11. DOI: 10.1016/j.ibmb.2013.10.009.

Wang, Z., Gerstein, M. and Snyder, M. (2009) RNA-Seq: a revolutionary tool for transcriptomics. *Nature Reviews Genetics* 10(1), 57–63. DOI: 10.1038/nrg2484.

Wang, B., Tseng, E., Regulski, M., Clark, T.A., Hon, T. *et al.* (2016) Unveiling the complexity of the maize transcriptome by single-molecule long-read sequencing. *Nature Communications* 7(1), 11708. DOI: 10.1038/ncomms11708.

Wang, K., Hong, W., Jiao, H. and Zhao, H. (2017) Transcriptome sequencing and phylogenetic analysis of four species of luminescent beetles. *Scientific Reports* 7(1), 1814. DOI: 10.1038/s41598-017-01835-9.

Wang, S. and Gribskov, M. (2017) Comprehensive evaluation of de novo transcriptome assembly programs and their effects on differential gene expression analysis. *Bioinformatics* 33, 327–333.

Wang, X., Fang, X., Yang, P., Jiang, X., Jiang, F. *et al.* (2014) The locust genome provides insight into Swarm formation and long-distance flight. *Nature Communications* 5(1), 2957. DOI: 10.1038/ncomms3957.

Weirather, J.L., de Cesare, M., Wang, Y., Piazza, P., Sebastiano, V. *et al.* (2017) Comprehensive comparison of Pacific biosciences and Oxford nanopore technologies and their applications to transcriptome analysis. *F1000Research* 6, 100. DOI: 10.12688/f1000research.10571.2.

Welboren, W.-J., van Driel, M.A., Janssen-Megens, E.M., van Heeringen, S.J., Sweep, F.C. *et al.* (2009) ChIP-Seq of ERα and RNA polymerase II defines genes differentially responding to ligands. *The EMBO Journal* 28(10), 1418–1428. DOI: 10.1038/emboj.2009.88.

Whitton, C., Daub, J., Thompson, M. and Blaxter, M. (2004) Expressed sequence tags: medium-throughput protocols. In: Melville, S.E. (ed.) *Methods in Molecular Biology*. Humana Press, Totowa, New Jersey, pp. 075–092.

Wiegmann, B.M., Trautwein, M.D., Kim, J.-W., Cassel, B.K., Bertone, M.A. *et al.* (2009) Single-copy nuclear genes resolve the phylogeny of the holometabolous insects. *BMC Biology* 7(1), 34. DOI: 10.1186/1741-7007-7-34.

Wu, T., Dhami, G.K. and Thompson, G.J. (2018) Soldier-biased gene expression in a subterranean termite implies functional specialization of the defensive caste. *Evolution and Development* 20(1), 3–16. DOI: 10.1111/ede.12243.

Ylla, G., Piulachs, M.-D. and Belles, X. (2018) Comparative transcriptomics in two extreme Neopterans reveals general trends in the evolution of modern insects. *IScience* 4, 164–179. DOI: 10.1016/j.isci.2018.05.017.

Yoshizawa, K. (2011) Monophyletic Polyneoptera recovered by wing base structure. *Systematic Entomology* 36(3), 377–394. DOI: 10.1111/j.1365-3113.2011.00572.x.

Zajac, K.B., Amendt,J., Verhoff, A.M. and Zehner, R. (2018) Dating pupae of the blow fly *Calliphora vicina Robineau–Desvoidy 1830 (Diptera: Calliphoridae)* for post mortem interval—estimation: validation of molecular age markers. *Genes* 9(3), 153. DOI: 10.3390/genes9030153.

Zera, A.J. (2003) The endocrine regulation of wing polymorphism in insects: state of the art, recent surprises, and future directions. *Integrative and Comparative Biology* 43(5), 607–616. DOI: 10.1093/icb/43.5.607.

Zera, A.J. and Larsen, A. (2001) The metabolic basis of life history variation: genetic and phenotypic differences in lipid reserves among life history morphs of the wing-polymorphic cricket, *Gryllus firmus*. *Journal of Insect Physiology* 47(10), 1147–1160. DOI: 10.1016/S0022-1910(01)00096-8.

Zera, A.J., Sall, J. and Grudzinski, K. (1997) Flight-muscle polymorphism in the cricket *Gryllus firmus*: muscle characteristics and their influence on the evolution of flightlessness. *Physiological Zoology* 70(5), 519–529. DOI: 10.1086/515865.

Zera, A.J. (2009) Wing polymorphism in Gryllus (Orthoptera: Gryllidae): proximate endocrine, energetic and biochemical mechanisms underlying morph specialization for flight vs. reproduction. In: Whitman, D.W. and Ananthakrishnan, T.N. (eds) *Phenotypic Plasticity of Insects: Mechanisms and Consequences*. Science Publishers, Inc., Enfield, New Hampshire, pp. 609–653.

Zhao, S., Fung-Leung, W.-P., Bittner, A., Ngo, K. and Liu, X. (2014) Comparison of RNA-Seq and microarray in transcriptome profiling of activated T cells. *PLoS ONE* 9(1), e78644. DOI: 10.1371/journal.pone.0078644.

Zhao, J.-Y., Zhao, X.-T., Sun, J.-T., Zou, L.-F., Yang, S.-X. *et al.* (2017) Transcriptome and proteome analyses reveal complex mechanisms of reproductive diapause in the two-spotted spider mite, *Tetranychus urticae*. *Insect Molecular Biology* 26(2), 215–232. DOI: 10.1111/imb.12286.

2 From Reads to Genes

CHENGRAN ZHOU,[1,2,3] GUANLIANG MENG,[1,2] AND SHANLIN LIU[1,2,4*]

[1]BGI-Shenzhen, Shenzhen, China;
[2]China National GeneBank, BGI-Shenzhen, Shenzhen, China;
[3]Key Laboratory of Bio-Resource and Eco-Environment of Ministry of Education, College of Life Sciences, Sichuan University, Chengdu, China;
[4]Beijing Advanced Innovation Center for Food Nutrition and Human Health, China Agricultural University, Beijing, China

2.1 Introduction

The burgeoning field of high-throughput sequencing (HTS) has enabled large-scale sequencing to be accessible for a vast range of organisms, thus significantly enhancing our understanding of the complexity of transcriptomes. With the increased sequencing capacities and reduced cost per base, transcriptomics is increasingly being utilized to investigate various scientific questions. Despite the broad scope of the scientific questions and interests of different studies, the applications of transcriptomics are mainly gene set construction, differential gene expression detection, and phylogenetic relation inference. This chapter introduces a brief analytical framework (Fig. 2.1) on which transcriptome studies are based: from millions of reads to biological meaningful genes.

2.2 Quality Control

HTS has enabled the generation of enormous amounts of data, in which the presence of low-quality bases, sequence artifacts, and adapter contamination would have a detrimental impact on the downstream analyses. The sequencing errors or adapter sequences can lead to mis-assemblies or being unable to map against reference genomes, and increase computer resource consumption. Therefore, preprocessing of the raw reads, including removal of adapter sequences, low-quality sequences, and reads with excessive ambiguous

*Corresponding author: shanlin1115@gmail.com

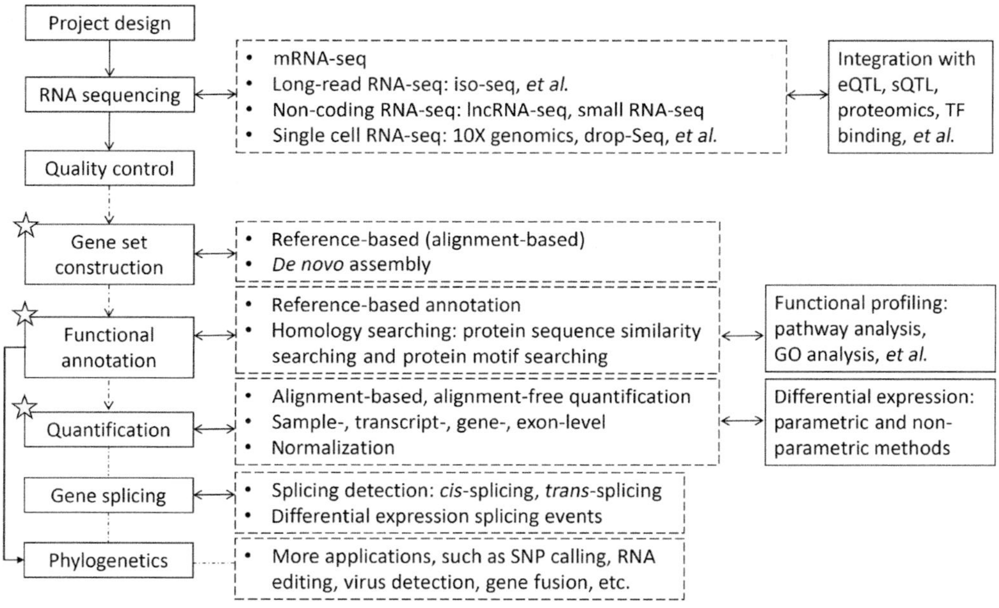

Fig. 2.1. Overview of the analytical framework of transcriptome studies, summarizing the key and possible steps over the broad spectrum of RNA-seq analysis.

bases ('N's), is a critical step to ensure data quality and prevent the substandard reads from subsequent analysis pipelines. Some commonly used tools to such ends include Trimmomatic (Bolger *et al.*, 2014), Cutadapt (Martin, 2011), NGS QC Toolkit (Patel and Jain, 2012), SOAPnuke (Chen *et al.*, 2017), Skewer (Jiang *et al.*, 2014), and FastQC (Andrews, 2010).

2.3 Gene Set Construction

The most commonly used HTS platforms, Illumina HiSeq and MGI DNBSEQ, generate short reads – paired-end or single-end (e.g. 2 × 100bp paired ends). To obtain transcripts, one needs assemblers to connect the short reads and reconstruct each transcript and gene. There are mainly two strategies used for transcriptome assembly: (i) reference-based assembly; and (ii) *de novo* assembly (Fig. 2.2) (Martin and Wang, 2011; Huang *et al.*, 2016; Li *et al.*, 2014).

2.3.1 Reference-based assembly

When a high-quality genome for the target transcriptome is available, one can try the reference-based assembly method, which mainly invokes three steps: (i) mapping reads to the reference genome with splice-aware aligners, such as HISAT2 (Kim *et al.*, 2015), TOPHAT2 (Kim *et al.*, 2013), or STAR (Dobin *et al.*, 2013); (ii) clustering the overlapping reads on each locus to build

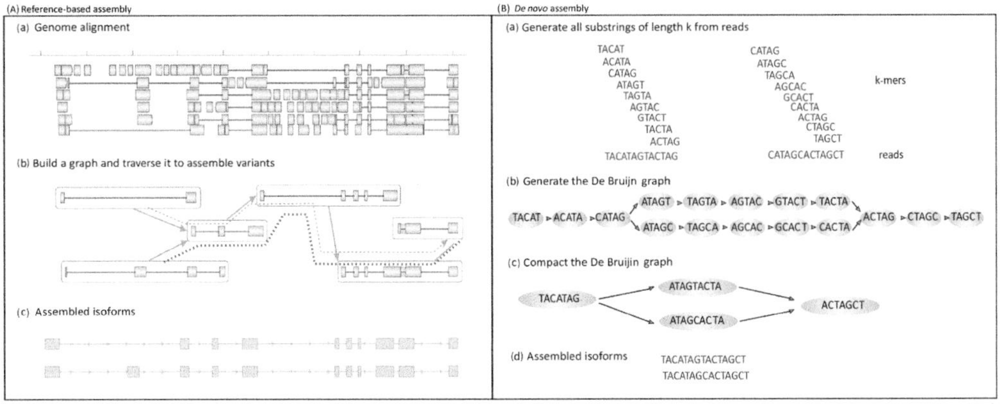

Modified from Martin, J. A., & Wang, Z. (2011).

Fig. 2.2. Overview of transcriptome assembly strategy. (A) The steps of the reference-based assembly method: (a) splice-align reads (gray blocks) to the reference; (b) construct a graph to represent all possible isoforms based on the alignment, and then generate alternative paths (gray and dark gray dashed lines) by traversing the graph; (c) construct isoforms. (B) The steps of the *de novo* assembly method: (a) generate k-mers (substrings of length k) from each read; (b) construct De Bruijn graph using k-mers (each k-mer represents a node and overlapped nodes are connected; single-nucleotide differences cause ‘bubbles’ in the De Bruijn graph); (c) compact chains of adjacent nodes into a single node and then construct isoforms using all paths through the graph; (d) generate final isoforms.

a graph representing all possible isoforms based on the alignment; and (iii) traversing the graph to generate individual isoforms. The latter two steps can be accomplished using software such as StringTie (Pertea *et al.*, 2015), Cufflinks (Trapnell *et al.*, 2012), etc.

There is an explosive growth of the number of available high-quality genomes and this growth pace is expected to increase rapidly as a result of the plunging cost of sequencing and the ongoing genome projects, such as the Earth BioGenome Project (Lewin *et al.*, 2018) and the 5000 Insect Genome Project (i5K Consortium, 2013). Therefore, one expects to see increasing application of reference-based assemblies, because they are affected little by contaminations and sequencing artifacts due to the filtering role of reference genome mapping. Also, reference-based assembly is sensitive and can recover low-abundance transcripts, which gives a chance to find novel transcripts that are not present in the current annotation. Moreover, compared with the *de novo* assembly strategy, reference-based requires less computer resources and requires fewer input reads, and, as a result, will make the sequencing budget more flexible.

2.3.2 *De novo* assembly

When the available genomes are phylogenetically distant from the target transcriptome or of low quality, the reference-based assembly method may lead to incomplete transcripts due to inaccurate reads-to-genome alignment. A different strategy is needed to make an alignment without a genome from the beginning, or *de novo*.

Most of the *de novo* assemblers adopt the De Bruijn graph (DBG) theory, in which substrings of reads are represented as nodes in the graph and transcripts are recovered by traversing the graph. The DBG is also commonly used in genome assembly. However, the genome assemblers assume that the sequencing depths along the whole genome are approximately the same, while those of different transcripts vary significantly. One also expects multiple isoforms on a single locus. Therefore, specific algorithms that take the aforementioned scenarios into account have been developed to construct transcripts from short reads, such as SOAPdenovo-trans (Xie *et al.*, 2014), Trinity (Grabherr *et al.*, 2011), Bridger (Chang *et al.*, 2015), IDBA-Tran (Peng *et al.*, 2013), Oases (Schulz *et al.*, 2012), Trans-ABySS (Robertson *et al.*, 2010), etc. The k-mer size that defines the sequence overlap between two contiguous reads has significant influence on the assembly quality for methods based on DBG. We recommend a large k-mer size, e.g. 51, when one primarily aims to achieve a more contiguous assembly and cares less about the loss of lowly expressed transcripts. On the other hand, small k-mer size is preferred in order to capture poorly expressed transcripts, resulting in more fragmented and diverse transcripts. Some studies tried to use multiple k-mers to generate multiple transcript sets separately, then combined the sets to form a final assembly using tools such as Rnnotator (Martin *et al.*, 2010), Multiple-k (Surget-Groba and Montoya-Burgos, 2010) and Trans-ABySS (Robertson *et al.*, 2010).

In practice, there are also studies combining the two assembly strategies either by align-then-assemble or by assemble-then-align (Martin and Wang, 2011). The align-then-assemble method has greater computer resource efficiency over the assemble-then-align method when the reference genome is of high quality, because by first mapping reads onto a reference genome, only a subset of reads will remain for the *de novo* assembly. However, when the reference genome quality is reduced (e.g. assembly errors, fragmented), one may want to perform the *de novo* assembly first, then align both the assemblies and the unassembled reads onto the reference genome to generate longer and more complete transcripts. This strategy takes advantage of the high sensitivity of reference-based assemblers while leveraging the ability of *de novo* assemblers to detect novel and trans-spliced transcripts.

2.3.3 Assembly quality control

It is critical to assess the quality of the assembled transcriptome sequences before any downstream analyses, especially for the assemblies obtained using a *de novo* method. One of the concerns is contamination, which can derive from sampling, sample transferring, laboratory experiments, or sequencing. One might want to filter out those sequences of high similarity between different taxa that pooled in the same sequencing lane, which could be cross-sample contaminations due to 'index hopping' (Kircher *et al.*, 2011), and those sequences of high similarity to the National Center for Biotechnology Information (NCBI) UniVec database (Kitts *et al.*, 2011). However, it is worth noting that a match to the cloning vectors or microbes does not necessarily mean that the identified sequence is a contaminant; instead, those transcriptome sequences can result from foreign genetic material incorporated into the genome (ex: horizontal gene transfer) or come from symbionts, which are themselves targets of study (Shi *et al.*, 2016).

2.4 Transcriptome Annotation

Transcriptome annotation refers to the process of protein function prediction and the related pathways of signaling and metabolism. The function annotation works following the well-known rule, 'sequence determines structure; structure determines function.' The protein structure determines its function, while the protein structure is determined by the sequence of amino acids, which is determined by the gene encoding it.

2.4.1 Functional annotation

Transcriptomes of species with high-quality annotated genomes, such as *Aedes aegypti* (Nene *et al.*, 2007), *Drosophila melanogaster* (Adams *et al.*, 2000), *Nasonia vitripennis* (Newton *et al.*, 2016), *Bombyx mori* (Xia *et al.*, 2004), etc., could be functionally annotated easily by mapping to the reference genome or existing gene set using genome alignment tools, such as HISAT (Kim *et al.*, 2015), Bowtie2 (Langmead and Salzberg, 2012) and STAR (Dobin *et al.*, 2013). For transcriptomes of species without well-annotated genome references and for novel transcripts, one conducts functional annotation by homology searching based on sequence similarity and protein motif structure.

Functional annotation can start either with assembled scaffolds or with coding sequences (CDS) that are predicted from the assemblies by tools like TransDecoder (Haas *et al.*, 2013). One searches for homologies for sequences via the publicly available protein sequence/motif databases, including the NCBI non-redundant (nr) protein database (Pruitt *et al.*, 2005), UniProtKB/Swiss-Prot (UniProt, 2019), the Kyoto Encyclopedia of Genes and Genomes (KEGG) pathway database (Kanehisa *et al.*, 2019), clusters of euKaryotic Orthologous Groups (KOG) (Tatusov *et al.*, 2003), database of Orthologous Groups (OrthoDB) (Kriventseva *et al.*, 2018), the Protein Families database (Pfam) (El-Gebali *et al.*, 2019), InterPro database (Mitchell *et al.*, 2019), or evolutionary genealogy of genes: Non-supervised Orthologous Groups (EggNOG) (Huerta-Cepas *et al.*, 2019); as well as some specific protein databases constructed from particular organisms or functional system-related proteins, e.g. Flybase (Thurmond *et al.*, 2019), InsectBase (Yin *et al.*, 2016), ChiloDB (Yin *et al.*, 2014), SilkDB (Duan *et al.*, 2010), ImmunoDB (Fenoglio and Giacomini, 2014), modENCODE (Gerstein *et al.*, 2010), and ASGARD (Zeng and Extavour, 2012). Although all these databases can be used for functional annotation, they contain different characteristics and some details about the most frequently used protein databases are summarized in Table 2.1.

Functional annotation in most cases is accomplished by homology searching to more than one database using the corresponding tools. For

Table 2.1. Frequently used protein databases

nr	The non-redundant protein database (nr) is a collection of sequences from several sources, including translations from annotated coding regions in GenBank, RefSeq and Third Party Annotation (TPA), as well as records from SwissProt, the Protein Information Resource (PIR), Protein Research Foundation (PRF), and Protein Data Bank (PDB). Protein sequences are the fundamental determinants of biological structure and function. ftp://ftp.ncbi.nlm.nih.gov/blast/db
KOG	EuKaryotic Orthologous Groups (KOG) is a eukaryote-specific version of the Clusters of Orthologous Groups (COG) tool for identifying ortholog and paralog proteins. ftp://ftp.ncbi.nih.gov/pub/COG/KOG
KEGG	KEGG (Kyoto Encyclopedia of Genes and Genomes) is a collection of databases utilized for bioinformatics research and education, including data analysis in genomics, metagenomics, metabolomics and other omics studies, modeling and simulation in systems biology, and translational research in drug development. http://www.genome.jp/kegg
SwissProt	UniProtKB/Swiss-Prot is the manually annotated and reviewed section of the UniProt Knowledgebase (UniProtKB). Thus, it is a database in which protein sequences were annotated with high quality by bringing together experimental results, computed features, and scientific conclusions. http://ftp.ebi.ac.uk/pub/databases/swissprot
InterPro	InterPro is a resource that provides functional analysis of protein sequences by classifying them into families and predicting the presence of domains and important sites. http://www.ebi.ac.uk/interpro
GO	The Gene Ontology (GO) project is a major bioinformatics initiative to develop a computational representation of our evolving knowledge of how genes encode biological functions at the molecular, cellular and tissue system levels. http://geneontology.org
AnimalTFDB	The Animal Transcription Factor DataBase (AnimalTFDB) is a comprehensive transcription factor (TF) database, in which all the genome-wide TFs in 50 sequenced animal genomes are identified and classified. AnimalTFDB provides annotation information for all the TFs, transcription co-factors and chromatin remodeling factors in the database. http://www.bioguo.org/AnimalTFDB/

example, one can use protein sequence databases (nr, KEGG, *et al.*) for homology identification using sequence similarity-based searching programs, such as Blastx (Altschul *et al.*, 1990) and Diamond (Buchfink *et al.*, 2015). One can also annotate sequences by domains and conservative families via databases such as Pfam (El-Gebali *et al.*, 2019) and orthoDB (Bateman *et al.*, 2004) using Hmmer (Finn *et al.*, 2011), RPS-BLAST (McGinnis and Madden, 2004), or OrthoMCL (Li *et al.*, 2003). Further annotations include searching the InterPro database using InterProScan (Quevillon *et al.*, 2005) and EggNOG database (Huerta-Cepas *et al.*, 2019) using EggNOG-mapper (Huerta-Cepas *et al.*, 2017). The Gene Ontology (GO) term of each transcript can be obtained using Blast2GO (Conesa *et al.*, 2005), GOstat (Beissbarth and Speed, 2004), or Annot8r (Schmid and Blaxter, 2008) based on nr annotation results. One can also predict transcription factors for each gene via the Animal Transcription Factor DataBase (AnimalTFDB) (Zhang *et al.*, 2011) using Hmmer (Finn *et al.*, 2011).

In addition to functional annotation based on the databases with proteins of known functions, for those transcripts that one cannot find homologs for, the candidate coding regions of each transcript can still be identified based on protein-coding regions identifiers, such as Transdecoder (Haas *et al.*, 2013), ESTscan (Iseli *et al.*, 1999), GeneMarkS-T (Tang *et al.*, 2015), Prodigal (Hyatt *et al.*, 2010), PhyloCSF (Lin *et al.*, 2011), and CPAT (Wang *et al.*, 2013a). The longest open reading frame (ORF) would represent the encoded genes as putative novel transcripts in most cases.

2.4.2 Cases

Crawford *et al.* (2010) constructed a set of 15,527 transcripts for *Anopheles funestus* via comparisons with four Dipteran transcriptomes as well as multiple functional and conserved protein domain databases using Blastx, TBlastx, RPS-BLAST, and Blast2GO; and identified 339 contigs as putative immune genes and 5434 transcripts as 1:1 orthologous between *An. funestus* and *An. gambiae*.

Geib *et al.* (2014) annotated transcripts by steps including: (i) BLASTP searching against the UniProtKB/Swiss-Prot database (UniProt, 2019); (ii) identifying protein domains through searching the Pfam_A database using HMMER; (iii) annotating signal peptides and transmembrane domains with SignalP and TMHMM, respectively; and (iv) loading the outputs into the database to assign each transcript the EggNOG and GO terms and exporting the final results as a delimited file.

Sim *et al.* (2015) constructed a gene set for *Bactrocera cucurbitae*, containing 10,425 unigenes and 17,654 transcript isoforms that account for about 70–80% of the protein-coding genes in the genome. In this study, the ORFs were predicted using Transdecoder and transcripts were annotated by searching UniProtKB/SwissProt, Pfam, InterPro, GO, and a *Drosophila* proteins specific database using InterProScan5 and BLASTP.

2.5 Differential Expression Analysis

2.5.1 Quantification

Differential expression (DE) analysis (Pradeep Kumar *et al.*, 2018) constitutes a large proportion of the studies associated with RNA-seq. It is used to identify the genes that are expressed differently under different conditions, as those genes are supposed to play critical roles in biological processes, metabolism regulation, or development (Oshlack *et al.*, 2010). Briefly, one can identify differentially expressed genes in three steps (Fig. 2.3). First, calculate and normalize the read counts for each gene or isoform; then identify the differential expressed genes using one or multiple statistical methods; and finally, validate the differentially expressed genes (DEG) using qRT-PCR.

To count the abundance of each transcript, one can either map the raw reads to a reference genome or use a pre-constructed gene set using an RNA alignment tool and then aggregate raw counts of mapped reads using programs such as HTSeq-count (Anders *et al.*, 2015), featureCounts (Liao *et al.*, 2014), and RSEM (Li and Dewey, 2011). One can also get the read count directly by using alignment-free methods for organisms that have well-curated full-transcript datasets (Table 2.2).

For the alignment-based method, several commonly used tools designed for spliced alignment are suitable, such as STAR (Dobin *et al.*, 2013), HISAT2 (Kim *et al.*, 2015), TOPHAT2 (Kim *et al.*, 2013), RSEM (Li and Dewey, 2011), and eXpress (Roberts and Pachter, 2013). It is worth noting that additional mapping quality control can further alleviate various bias issues generated during the sample preparation or sequencing. For instance, parameters such

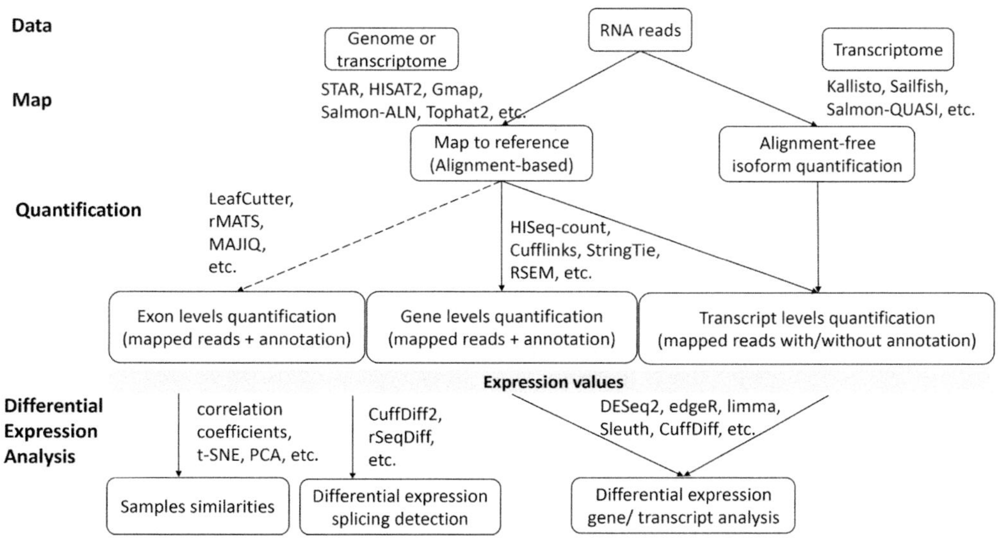

Fig. 2.3. Overview of the RNA-seq pipeline for differential expression analysis. Text in each row presents the widely used methods and software examples for each step.

Table 2.2. Overview of RNA aligners.

Alignment-free methods	
Salmon	Salmon implements a two-phase inference procedure including both online and offline iterations of Expectation-Maximization (EM). Salmon is a flexible tool that has two modes of quantification (Zhang *et al.*, 2017).https://combine-lab.github.io/salmon/
Kallisto	Kalllisto introduces a De Bruijn graph to achieve 'pseudo-alignment' by checking the compatibility between short reads with transcripts.https://pachterlab.github.io/kallisto
Sailfish	Sailfish implements a k-mer approach, replacing traditional read alignment with the allocation of exact k-mers to transcripts, to quantify gene or isoform abundance.http://www.cs.cmu.edu/~ckingsf/software/sailfish/
Alignment-based methods	
STAR	The Spliced Transcripts Alignment to a Reference (STAR) software is based on an alignment algorithm that uses sequential maximum mappable seed search in uncompressed suffix arrays followed by seed clustering and stitching procedures.https://github.com/alexdobin/STAR
HISAT2	Implements a new indexing scheme called Hierarchical Graph FM index (HGFM) that enables effective alignment of sequencing reads.https://ccb.jhu.edu/software/hisat2/
RSEM	The RSEM package supports threads for parallel computation of the EM algorithm, single-end and paired-end read data, quality scores, variable-length reads, and RSPD estimation.https://github.com/deweylab/RSEM
eXpress	A streaming tool based on an online-EM algorithm for quantifying the abundances of a set of target sequences from sampled subsequences.https://pachterlab.github.io/eXpress/overview.html

as the GC content, the percentage of mapped reads, and the uniformity of read depth on exons and the mapped strand, can be checked using mapping quality control tools like Picard (McKenna *et al.*, 2010), ReSeqTools (Jia *et al.*, 2013), and RSeQC (Wang *et al.*, 2012). The obtained raw read counts are not ready for gene expression comparison between samples, since various factors including the gene length, library size, and sequencing depth contribute to the final value of gene expression level, so one needs to take all these factors into account and remove the bias using a normalization method, such as Total Count (TC), Upper Quartile, Median, Quantile, Relative Log Expression (RLE), the DESeq normalization, Trimmed mean of M values (TMM), the transcripts per million mapped reads (TPM), and the Reads Per Kilobase per Million mapped reads (RPKM) normalizations (Conesa *et al.*, 2016). The TPM estimation can eliminate statistical biases inherent in the RPKM and is regarded as a better unit than RPKM (Wagner *et al.*, 2012).

The alignment-free approaches use k-mer to assign reads to their corresponding transcript and will produce normalized gene expression values directly. Therefore, compared with the alignment-based methods, they are more computationally efficient. However, for lowly abundant and small-size RNAs they show less reliable performance (Wu *et al.*, 2018).

2.5.2 Expression analysis

Before DE analysis, reproducibility checking among replicates can avoid error outliers generated due to maloperations during the procedures of sample preparation or sequencing. Approaches including correlation coefficients (Arbeitman *et al.*, 2002), alternative transcriptome overlap measure (TROM), partial correlation, partial mutual information (PMI), etc., can be used to measure sample similarities in various scenarios. Dimension reduction and visualization of the high-dimensional gene expression data, such as principal component analysis (PCA), t-stochastic neighbor embedding (t-SNE), etc. (Li and Li, 2018), can also be utilized to check the similarities.

After calculating gene abundance, several widely used tools like DESeq2 (Love *et al.*, 2014), baySeq (Hardcastle and Kelly, 2010) and edgeR (Robinson *et al.*, 2010) were developed to detect DEGs using a negative binomial (NB) distribution model. By contrast, methods of limma (Ritchie *et al.*, 2015), voom (Law *et al.*, 2014), Sleuth (Pimentel *et al.*, 2017), Cuffdiff (Trapnell *et al.*, 2012), and Ballgown (Frazee *et al.*, 2015) were developed based on a linear model. There are also several non-parametric-based tools, for example, NOIseq (Tarazona *et al.*, 2015), SAMseq (Li and Tibshirani, 2013), etc. Parametric methods are applied to differential gene expression assuming the transcript expression can be mapped into a particular distribution, such as Poisson or NB, while non-parametric methods will not define the data distribution to a finite set of parameters (Costa-Silva *et al.*, 2017). Currently, the NB distribution with generalized linear models are the most widely used algorithms to determine significantly expressed genes, because the DEG studies usually contain limited sample numbers and therefore make it hard to accurately estimate each gene's expression variance, which is critical to determine whether the changes are caused by chance or not. The typical strategy is to divide gene sets into certain groups according to their expression level and then estimate a global variance for each group. After that, these variance parameters are applied to model the NB distribution for each gene. Since the NB, compared with Poisson, does not presume an equal value for the mean and variance, it can therefore generate a more accurate estimation of variances for each gene.

The results including log2 fold-change, adjusted P-value, and false discovery rates (FDR) can be used to identify significant DEGs and can be visualized using stand-alone tools like CummeRbund (Goff *et al.*, 2012) and Sashimi plots (Katz *et al.*, 2013), or tools in integrated packages, such as DESeq2 (Love *et al.*, 2014), DEXseq (Bonin *et al.*, 2015), etc.

In addition to the regulation of gene expression level, RNA splicing plays a vital role in genetics by increasing mRNA diversity and protein diversities from a limited number of genes in insects. For instance, instead of the well-known expression quantitative trait locus (eQTL), which refers to SNPs that may influence the expression levels of genes, the genetic variation that affects splicing regulation, referred to as splicing quantitative trait loci (sQTLs), can lead to differences in alternative splicing between individuals, consequently influencing traits like disease susceptibility and drug response (Tian *et al.*, 2018). Methods have been developed to identify differentially splicing genes (DSGs) between samples at both the isoform resolution and the exon resolution, of which the isoform-based method detects changes in the proportion of each isoform within the total gene expression (e.g. CuffDiff2 (Trapnell *et al.*, 2013) and rSeqDiff (Shi and Jiang, 2013)), while the exon-based approach detects signals of alternative splicing by directly comparing the distributions of reads on exons and junctions (e.g. DEXseq (Bonin *et al.*, 2015), DSGSeq *(Wang et al., 2013b)*, rMATS (Shen *et al.*, 2014), rDiff (Drewe *et al.*, 2013) and DiffSplice (Hu *et al.*, 2012)). The exon-based approaches are able to deliver more accurate results in comparison with the isoform-based methods (Conesa *et al.*, 2016). However, both strategies rely on specific statistical models or pre-defined splicing events. The predefined transcript models arguably are particularly limiting in comparison studies, as aberrant transcripts may be case specific: for example, particular splicing events only happened in a subset of individuals of the disease group (Xiong *et al.*, 2015). Furthermore, the non-spliced reads contributing to abundance estimation of isoforms or exons are often ambiguous, due to substantial overlaps between different isoforms. Therefore, alternative approaches focusing on only the spliced reads and particularly recording what is removed (intron excision) in each splicing event were developed to identify complex splicing events, such as MAJIQ (Vaquero-Garcia *et al.*, 2016) and LeafCutter (Li *et al.*, 2018).

2.5.3 Cases

RNA-Seq has been applied to profile expression in species representing a wide array of insect sociality, including in advanced eusocial honey bees (Chen *et al.*, 2012), an intermediately social bumble bee (Harrison *et al.*, 2015), three primitively eusocial species of *Polistes* wasps (Standage *et al.*, 2016; Berens *et al.*, 2014), and ants (Ometto *et al.*, 2010; Simola *et al.*, 2013; Morandin *et al.*, 2015). Differences in gene expression and alternative splicing between castes of social insects provide a striking example of how environment induces phenotypic plasticity (Price *et al.*, 2018).

Lin *et al.* (2016) sequenced the mRNA of 726 individuals from the *Drosophila* Genetic Reference Panel and found that the best analytical approach to this data was to normalize the read counts using the DESeq method and apply a generalized linear model assuming the NB distribution using either edgeR or DESeq.

Teets *et al.* (2012) quantified the transcriptional mechanisms of extreme dehydration tolerance of *Belgica antarctica*, determined differentially expressed genes using DESeq, performed hierarchical clustering using hclust, and identified enriched GO terms using GOsEq with P-values corrected using the Benjamini and Hochberg method. Gene expression changes in response to dehydration indicated an up-regulation of cellular recycling pathways.

Duan *et al.* (2017) sequenced and *de novo* assembled the transcriptomes for *Mythimna separate* exposed to four different light conditions. DEGs were determined with a threshold of FDR ≤ 0.001 and an absolute value of log2Ratio ≥ 1.

Graveley *et al.* (2011) explored the *Drosophila melanogaster* transcriptome in 30 distinct developmental stages using data from RNA-Seq, tiling microarrays, and cDNA sequencing, and reported 11,457 intron retention events, 3548 alternative 3′-splice sites, 3457 alternative 5′-splice sites, 1844 cassette exons, 831 multiple cassettes events, and 145 mutually exclusive splicing events that shift significantly between different cell types.

Qu *et al.* (2017) combined both isoform- and exon-based methods to search for lead (Pb) responsive sQTLs in *Drosophila melanogaster* using RNA-seq data and resulted in a total of 1236 Pb-responsive sQTLs.

2.6 Other Applications

The transcriptome can be utilized to unravel the evolution of insects. The 1KITE (1K Insect Transcriptome Evolution) (http://www.1kite.org/) project aims to study the transcriptomes of more than 1000 insect species, encompassing all recognized insect orders, and therefore could infer robust phylogenetic backbone trees for insects. The first symbolic achievement of the 1KITE project studied the transcriptomes of more than 100 insect species representing all extant insect orders and was able to produce statistically robust and congruent results from 1478 protein-coding genes, resolving previously controversial phylogenetic relationships (Misof *et al.*, 2014). In brief, it firstly identified 1478 single-copy nuclear genes present in all the reference species' official gene sets (14 published genomes in this work) by comparative genome analysis. Functional annotation of these genes revealed that many serve basic cellular functions and thus should be expressed in all stages of life and various tissue types. On the above assumption, the study was able to identify, on average, 98% of these genes in the 103 *de novo* sequenced transcriptomes with a graph-based approach using the best reciprocal genome- and transcriptome-wide hit criteria. Finally, the identified orthologous genes were utilized to build a nucleotide super-matrix and infer the phylogenetic tree using a method based on maximum likelihood (ML).

The following studies applied the same strategies for different insect groups to: (i) infer the phylogeny and divergence times of all major lineages of Hymenoptera by analyzing 3256 protein-coding genes in 173 insect species (Peters *et al.*, 2017); (ii) resolve the relationships among major hemipteroid lineages by analyzing a taxonomically comprehensive dataset comprising sequences of 2395 single-copy, protein-coding genes for 193 samples of hemipteroid insects and outgroups (Johnson *et al.*, 2018); and (iii) provide a formal reconstruction of the evolutionary history of both Polyneoptera and early winged insects by combining phylogenetic analyses of a large transcriptomic dataset comprising 3014 protein-coding genes sampled from a total of 106 extant insect species, with a critical re-evaluation of morphological and embryological arguments for all recovered inter-ordinal nodes (Wipfler *et al.*, 2019). With the onset of HTS, transcriptome-based phylogenetic studies can take advantage of thousands of single-copy protein-coding genes with a low budget to resolve the evolutionary history for lineages of phenotypic or biodiversity interest, and are continuing to shed new light on our understanding of the evolution of trait gain and loss, various forms of social behavior, changes in lifestyle, and associated morphological specializations.

Other applications of transcriptomics in insects include but are not limited to: (i) studying single nucleotide polymorphisms (SNPs) between samples to test the association of the identified SNPs with a phenotype of interest (Lopez-Maestre *et al.*, 2016); (ii) studying the RNA editing (mainly the A-to-I editing) profiles between different tissues or phenotypes to investigate the genes of different RNA editing levels and their functions (Yu *et al.*, 2016; Li *et al.*, 2014); (iii) detecting known and novel insect RNA viruses, observing viral mutagenesis and recombination, and understanding virus-host interactions (Li *et al.*, 2015; Shi *et al.*, 2016; Zhang *et al.*, 2018; Zhou *et al.*, 2018); and (iv) identifying chimeric mRNAs generated by gene fusion in *cis* or in *trans* (Gabler *et al.*, 2005; McManus *et al.*, 2010; Kong *et al.*, 2015).

2.7 Perspective

Transcriptome datasets are one of the most preponderant resources, most of which were generated using short-read sequencing platforms. However, short-read sequencing has inherent limitations such as GC bias, difficulties mapping to repetitive elements, etc. (Alkan *et al.*, 2010; Ozsolak and Milos, 2011). Long-read single molecule sequencers, such as the currently commercially available PacBio single-molecule real-time (SMRT) sequencing technology (Chin *et al.*, 2013) and Oxford Nanopore Technologies (ONT) (Laver *et al.*, 2015), can resolve these obstacles. The long-read sequencer allows scientists to generate ultra-long reads covering full transcripts, and can achieve full-length cDNA sequences without assembly and more accurately characterize transcript isoforms (Jain *et al.*, 2018). For example, scientists applied SMRT sequencing for six tissue types from maize, generating 111,151 transcript isoform sequences (almost double the number of transcripts documented in the former reference genome annotation) associated with nearly 27,000 genes, representing 70% of previously annotated maize genes as well as novel isoforms and even some novel genes (Wang *et al.*, 2016). Both the long-read sequencers, since becoming commercially available, have evolved into new systems with significantly higher throughput: SMRT into RSII and Sequel systems, ONT into the PromethION system. These developments bring the cost per base down to an accessible level, offering us an excellent opportunity to update and accumulate gene sets from various species of all domains of life.

References

Adams, M.D., Celniker, S.E., Holt, R.A., Evans, C.A., Gocayne, J.D. (2000) The genome sequence of *Drosophila melanogaster*. *Science* 287(5461), 2185–2195. DOI: 10.1126/science.287.5461.2185.

Alkan, C., Sajjadian, S. and Eichler, E.E. (2010) Limitations of next-generation genome sequence assembly. *Nature Methods* 8(1), 61–65. DOI: 10.1038/nmeth.1527.

Altschul, S.F., Gish, W., Miller, W., Myers, E.W. and Lipman, D.J. (1990) Basic local alignment search tool. *Journal of Molecular Biology* 215(3), 403–410. DOI: 10.1016/S0022-2836(05)80360-2.

Anders, S., Pyl, P.T. and Huber, W. (2015) HTSeq--a Python framework to work with high-throughput sequencing data. *Bioinformatics* 31(2), 166–169. DOI: 10.1093/bioinformatics/btu638.

Andrews, S. (2010) FastQC: a quality control tool for high throughput sequence data. Cambridge, UK, Babraham Institute. Available at: http://www.bioinformatics.babraham.ac.uk/projects/fastqc (accessed 8 June 2019).

Arbeitman, M.N., Furlong, E.E., Imam, F., Johnson, E., Null, B.H. (2002) Gene expression during the life cycle of *Drosophila melanogaster*. *Science* 297(5590), 2270–2275. DOI: 10.1126/science.1072152.

Bateman, A., Coin, L., Durbin, R., Finn, R.D., Hollich, V. (2004) The Pfam protein families database. *Nucleic Acids Research* 32(90001), D138–D141. DOI: 10.1093/nar/gkh121.

Beissbarth, T. and Speed, T.P. (2004) GOstat: find statistically overrepresented Gene Ontologies within a group of genes. *Bioinformatics* 20(9), 1464–1465. DOI: 10.1093/bioinformatics/bth088.

Berens, A.J., Hunt, J.H. and Toth, A.L. (2014) Comparative transcriptomics of convergent evolution: different genes but conserved pathways underlie caste phenotypes across lineages of eusocial insects. *Molecular Biology and Evolution* 32(3), 690–703. DOI: 10.1093/molbev/msu330.

Bolger, A.M., Lohse, M. and Usadel, B. (2014) Trimmomatic: a flexible trimmer for Illumina sequence data. *Bioinformatics* 30(15), 2114–2120. DOI: 10.1093/bioinformatics/btu170.

Bonin, M.C., Boström-Einarsson, L., Munday, P.L. and Jones, G.P. (2015) The prevalence and importance of competition among coral reef fishes. *Annual Review of Ecology, Evolution, and Systematics* 46(1), 169–190. DOI: 10.1146/annurev-ecolsys-112414-054413.

Buchfink, B., Xie, C. and Huson, D.H. (2015) Fast and sensitive protein alignment using DIAMOND. *Nature Methods* 12(1), 59–60. DOI: 10.1038/nmeth.3176.

Chang, Z., Li, G., Liu, J., Zhang, Y., Ashby, C. *et al.* (2015) Bridger: a new framework for de novo transcriptome assembly using RNA-seq data. *Genome Biology* 16(1), 30. DOI: 10.1186/s13059-015-0596-2.

Chen, Y., Chen, Y., Shi, C., Huang, Z., Zhang, Y. *et al.* (2017) SOAPnuke: a MapReduce acceleration-supported software for integrated quality control and preprocessing of high-throughput sequencing data. *GigaScience* 7, gix120.

Chen, X., Hu, Y., Zheng, H., Cao, L., Niu, D. *et al.* (2012) Transcriptome comparison between honey bee queen- and worker-destined larvae. *Insect Biochemistry and Molecular Biology* 42(9), 665–673. DOI: 10.1016/j.ibmb.2012.05.004.

Chin, C.-S., Alexander, D.H., Marks, P., Klammer, A.A., Drake, J. *et al.* (2013) Nonhybrid, finished microbial genome assemblies from long-read SMRT sequencing data. *Nature Methods* 10(6), 563–569. DOI: 10.1038/nmeth.2474.

Conesa, A., Götz, S., García-Gómez, J.M., Terol, J., Talón, M. *et al.* (2005) Blast2GO: a universal tool for annotation, visualization and analysis in functional genomics research. *Bioinformatics* 21(18), 3674–3676. DOI: 10.1093/bioinformatics/bti610.

Conesa, A., Madrigal, P., Tarazona, S., Gomez-Cabrero, D., Cervera, A. *et al.* (2016) A survey of best practices for RNA-seq data analysis. *Genome Biology* 17(1), 13. DOI: 10.1186/s13059-016-0881-8.

Costa-Silva, J., Domingues, D. and Lopes, F.M. (2017) RNA-Seq differential expression analysis: an extended review and a software tool. *PloS ONE* 12(12), e0190152. DOI: 10.1371/journal.pone.0190152.

Crawford, J.E., Guelbeogo, W.M., Sanou, A., Traoré, A., Vernick, K.D. *et al.* (2010) De novo transcriptome sequencing in *Anopheles funestus* using Illumina RNA-seq technology. *PLoS ONE* 5(12), e14202. DOI: 10.1371/journal.pone.0014202.

Dobin, A., Davis, C.A., Schlesinger, F., Drenkow, J., Zaleski, C. *et al.* (2013) STAR: ultrafast universal RNA-seq aligner. *Bioinformatics* 29(1), 15–21. DOI: 10.1093/bioinformatics/bts635.

Drewe, P., Stegle, O., Hartmann, L., Kahles, A., Bohnert, R. *et al.* (2013) Accurate detection of differential RNA processing. *Nucleic Acids Research* 41(10), 5189–5198. DOI: 10.1093/nar/gkt211.

Duan, Y., Gong, Z., Wu, R., Miao, J., Jiang, Y. *et al.* (2017) Transcriptome analysis of molecular mechanisms responsible for light-stress response in *Mythimna separata* (Walker). *Scientific Reports* 7(1), 45188. DOI: 10.1038/srep45188.

Duan, J., Li, R., Cheng, D., Fan, W., Zha, X. *et al.* (2010) SilkDB v2.0: a platform for silkworm (*Bombyx mori*) genome biology. *Nucleic Acids Research* 38(suppl_1), D453–D456. DOI: 10.1093/nar/gkp801.

El-Gebali, S., Mistry, J., Bateman, A., Eddy, S.R., Luciani, A. *et al.* (2019) The Pfam protein families database in 2019. *Nucleic Acids Research* 47(D1), D427–D432. DOI: 10.1093/nar/gky995.

Fenoglio, D. and Giacomini, M. (2014) ImmunoDB: a web based tool to analyze preclinical data. *EHealth For Continuity of Care. Proceedings of MIE2014* 205, 438.

Finn, R.D., Clements, J. and Eddy, S.R. (2011) HMMER web server: interactive sequence similarity searching. *Nucleic Acids Research* 39(suppl), W29–W37. DOI: 10.1093/nar/gkr367.

Frazee, A.C., Pertea, G., Jaffe, A.E., Langmead, B., Salzberg, S.L. *et al.* (2015) Ballgown bridges the gap between transcriptome assembly and expression analysis. *Nature Biotechnology* 33(3), 243–246. DOI: 10.1038/nbt.3172.

Gabler, M., Volkmar, M., Weinlich, S., Herbst, A., Dobberthien, P. *et al.* (2005) Trans -splicing of the *mod (mdg4)* complex locus is conserved between the distantly related species *Drosophila melanogaster* and *D. virilis*. *Genetics* 169(2), 723–736. DOI: 10.1534/genetics.103.020842.

Geib, S.M., Calla, B., Hall, B., Hou, S. and Manoukis, N.C. (2014) Characterizing the developmental transcriptome of the oriental fruit fly, *Bactrocera dorsalis* (Diptera: Tephritidae) through comparative genomic analysis with *Drosophila melanogaster* utilizing modENCODE datasets. *BMC Genomics* 15(1), 942. DOI: 10.1186/1471-2164-15-942.

Gerstein, M.B., Lu, Z.J., Van Nostrand, E.L., Cheng, C., Arshinoff, B.I. *et al.* (2010) Integrative analysis of the *Caenorhabditis elegans* genome by the modENCODE project. *Science* 330(6012), 1775–1787. DOI: 10.1126/science.1196914.

Goff, L.A., Trapnell, C. and Kelley, D. (2012) CummeRbund: visualization and exploration of Cufflinks high-throughput sequencing data. *R package version 2*.

Grabherr, M.G., Haas, B.J., Yassour, M., Levin, J.Z., Thompson, D.A. *et al.* (2011) Full-length transcriptome assembly from RNA-Seq data without a reference genome. *Nature Biotechnology* 29(7), 644–652. DOI: 10.1038/nbt.1883.

Graveley, B.R., Brooks, A.N., Carlson, J.W., Duff, M.O., Landolin, J.M. *et al.* (2011) The developmental transcriptome of *Drosophila melanogaster*. *Nature* 471(7339), 473–479. DOI: 10.1038/nature09715.

Haas, B.J., Papanicolaou, A., Yassour, M., Grabherr, M., Blood, P.D. *et al.* (2013) De novo transcript sequence reconstruction from RNA-seq using the Trinity platform for reference generation and analysis. *Nature Protocols* 8(8), 1494–1512. DOI: 10.1038/nprot.2013.084.

Hardcastle, T.J. and Kelly, K.A. (2010) baySeq: empirical Bayesian methods for identifying differential expression in sequence count data. *BMC Bioinformatics* 11(1), 422. DOI: 10.1186/1471-2105-11-422.

Harrison, M.C., Hammond, R.L. and Mallon, E.B. (2015) Reproductive workers show queenlike gene expression in an intermediately eusocial insect, the buff-tailed bumble bee *Bombus terrestris*. *Molecular Ecology* 24(12), 3043–3063. DOI: 10.1111/mec.13215.

Hu, Y., Huang, Y., Du, Y., Orellana, C.F., Singh, D. *et al.* (2012) DiffSplice: the genome-wide detection of differential splicing events with RNA-seq. *Nucleic Acids Research* 41(2), e39–e39. DOI: 10.1093/nar/gks1026.

Huang, X., Chen, X.-G. and Armbruster, P.A. (2016) Comparative performance of transcriptome assembly methods for non-model organisms. *BMC Genomics* 17(1), 523. DOI: 10.1186/s12864-016-2923-8.

Huerta-Cepas, J., Forslund, K., Coelho, L.P., Szklarczyk, D., Jensen, L.J. *et al.* (2017) Fast genome-wide functional annotation through orthology assignment by eggNOG-mapper. *Molecular Biology and Evolution* 34(8), 2115–2122. DOI: 10.1093/molbev/msx148.

Huerta-Cepas, J., Szklarczyk, D., Heller, D., Hernández-Plaza, A., Forslund, S.K. *et al.* (2019) eggNOG 5.0: a hierarchical, functionally and phylogenetically annotated orthology resource based on 5090 organisms and 2502 viruses. *Nucleic Acids Research* 47(D1), D309–D314. DOI: 10.1093/nar/gky1085.

Hyatt, D., Chen, G.-L., LoCascio, P.F., Land, M.L., Larimer, F.W. *et al.* (2010) Prodigal: prokaryotic gene recognition and translation initiation site identification. *BMC Bioinformatics* 11(1), 119. DOI: 10.1186/1471-2105-11-119.

i5K Consortium (2013) The i5K initiative: advancing arthropod genomics for knowledge, human health, agriculture, and the environment. *Journal of Heredity* 104(5), 595–600. DOI: 10.1093/jhered/est050.

Iseli, C., Jongeneel, C.V. and Bucher, P. (1999) ESTScan: a program for detecting, evaluating, and reconstructing potential coding regions in EST sequences. *Proceedings. International Conference on Intelligent Systems for Molecular Biology* 99, 138–148.

Jain, M., Koren, S., Miga, K.H., Quick, J., Rand, A.C. *et al.* (2018) Nanopore sequencing and assembly of a human genome with ultra-long reads. *Nature Biotechnology* 36(4), 338–345. DOI: 10.1038/nbt.4060.

Jia, W., Qiu, K., He, M., Song, P., Zhou, Q. *et al.* (2013) SOAPfuse: an algorithm for identifying fusion transcripts from paired-end RNA-Seq data. *Genome Biology* 14(2), R12. DOI: 10.1186/gb-2013-14-2-r12.

Jiang, H., Lei, R., Ding, S.-W. and Zhu, S. (2014) Skewer: a fast and accurate adapter trimmer for next-generation sequencing paired-end reads. *BMC Bioinformatics* 15(1), 182. DOI: 10.1186/1471-2105-15-182.

Johnson, K.P., Dietrich, C.H., Friedrich, F., Beutel, R.G., Wipfler, B. *et al.* (2018) Phylogenomics and the evolution of hemipteroid insects. *Proceedings of the National Academy of Sciences* 115(50), 12775–12780. DOI: 10.1073/pnas.1815820115.

Kanehisa, M., Sato, Y., Furumichi, M., Morishima, K. and Tanabe, M. (2019) New approach for understanding genome variations in KEGG. *Nucleic Acids Research* 47(D1), D590–D595. DOI: 10.1093/nar/gky962.

Katz, Y., Wang, E.T., Silterra, J., Schwartz, S., Wong, B. *et al.* (2013) Sashimi plots: quantitative visualization of RNA sequencing read alignments. *arXiv: Genomics. arXiv* 3466, 1306.

Kim, D., Pertea, G., Trapnell, C., Pimentel, H., Kelley, R. *et al.* (2013) TopHat2: accurate alignment of transcriptomes in the presence of insertions, deletions and gene fusions. *Genome Biology* 14(4), R36. DOI: 10.1186/gb-2013-14-4-r36.

Kim, D., Langmead, B. and Salzberg, S.L. (2015) HISAT: a fast spliced aligner with low memory requirements. *Nature Methods* 12(4), 357–360. DOI: 10.1038/nmeth.3317.

Kircher, M., Sawyer, S. and Meyer, M. (2011) Double indexing overcomes inaccuracies in multiplex sequencing on the Illumina platform. *Nucleic Acids Research* 40(1), e3–e3. DOI: 10.1093/nar/gkr771.

Kitts, P., Madden, T., Sicotte, H., Black, L. and Ostell, J. (2011) UniVec Database. Available at: www.ncbi.nlm.nih.gov/VecScreen/UniVec.html (accessed 9 June 2019).

Kong, Y., Zhou, H., Yu, Y., Chen, L., Hao, P. *et al.* (2015) The evolutionary landscape of intergenic trans-splicing events in insects. *Nature Communications* 6(1), 8734. DOI: 10.1038/ncomms9734.

Kriventseva, E.V., Kuznetsov, D., Tegenfeldt, F., Manni, M., Dias, R. *et al.* (2018) OrthoDB v10: sampling the diversity of animal, plant, fungal, protist, bacterial and viral genomes for evolutionary and functional annotations of orthologs. *Nucleic Acids Research* 47(D1), D807–D811. DOI: 10.1093/nar/gky1053.

Langmead, B. and Salzberg, S.L. (2012) Fast gapped-read alignment with Bowtie 2. *Nature Methods* 9(4), 357–359. DOI: 10.1038/nmeth.1923.

Laver, T., Harrison, J., O'Neill, P.A., Moore, K., Farbos, A. *et al.* (2015) Assessing the performance of the Oxford nanopore technologies MinION. *Biomolecular Detection and Quantification* 3, 1–8. DOI: 10.1016/j.bdq.2015.02.001.

Law, C.W., Chen, Y., Shi, W. and Smyth, G.K. (2014) voom: precision weights unlock linear model analysis tools for RNA-seq read counts. *Genome Biology* 15(2), 1–17. DOI: 10.1186/gb-2014-15-2-r29.

Lewin, H.A., Robinson, G.E., Kress, W.J., Baker, W.J., Coddington, J. *et al.* (2018) Earth BioGenome Project: sequencing life for the future of life. *Proceedings of the National Academy of Sciences* 115(17), 4325–4333. DOI: 10.1073/pnas.1720115115.

Li, B. and Dewey, C.N. (2011) RSEM: accurate transcript quantification from RNA-Seq data with or without a reference genome. *BMC Bioinformatics* 12(1), 323. DOI: 10.1186/1471-2105-12-323.

Li, Y.I., Knowles, D.A., Humphrey, J., Barbeira, A.N., Dickinson, S.P. *et al.* (2018) Annotation-free quantification of RNA splicing using LeafCutter. *Nature Genetics* 50(1), 151–158. DOI: 10.1038/s41588-017-0004-9.

Li, W.V. and Li, J.J. (2018) Modeling and analysis of RNA-seq data: a review from a statistical perspective. *arXiv. Genomics* 6, 195–209.

Li, B., Fillmore, N., Bai, Y., Collins, M., Thomson, J.A. *et al.* (2014) Evaluation of de novo transcriptome assemblies from RNA-Seq data. *Genome Biology* 15(12), 553–553. DOI: 10.1186/s13059-014-0553-5.

Li, C.-X., Shi, M., Tian, J.-H., Lin, X.-D., Kang, Y.-J. *et al.* (2015) Unprecedented genomic diversity of RNA viruses in arthropods reveals the ancestry of negative-sense RNA viruses. *elife* 4, e05378. DOI: 10.7554/eLife.05378.

Li, L., Stoeckert, C.J. and Roos, D.S. (2003) OrthoMCL: identification of ortholog groups for eukaryotic genomes. *Genome Research* 13(9), 2178–2189. DOI: 10.1101/gr.1224503.

Li, J. and Tibshirani, R. (2013) Finding consistent patterns: a nonparametric approach for identifying differential expression in RNA-Seq data. *Statistical Methods in Medical Research* 22(5), 519–536. DOI: 10.1177/0962280211428386.

Li, Q., Wang, Z., Lian, J., Schiøtt, M., Jin, L. *et al.* (2014) Caste-specific RNA editomes in the leaf-cutting ant *Acromyrmex echinatior*. *Nature Communications* 5(1), 4943. DOI: 10.1038/ncomms5943.

Liao, Y., Smyth, G.K. and Shi, W. (2014) featureCounts: an efficient general purpose program for assigning sequence reads to genomic features. *Bioinformatics* 30(7), 923–930. DOI: 10.1093/bioinformatics/btt656.

Lin, Y., Golovnina, K., Chen, Z.-X., Lee, H.N., Negron, Y.L.S. *et al.* (2016) Comparison of normalization and differential expression analyses using RNA-Seq data from 726 individual *Drosophila melanogaster*. *BMC Genomics* 17(1), 28. DOI: 10.1186/s12864-015-2353-z.

Lin, M.F., Jungreis, I. and Kellis, M. (2011) PhyloCSF: a comparative genomics method to distinguish protein coding and non-coding regions. *Bioinformatics* 27(13), i275–i282. DOI: 10.1093/bioinformatics/btr209.

Lopez-Maestre, H., Brinza, L., Marchet, C., Kielbassa, J., Bastien, S. *et al.* (2016) SNP calling from RNA-seq data without a reference genome: identification, quantification, differential analysis and impact on the protein sequence. *Nucleic Acids Research* 44(Suppl 6), e148–e148. DOI: 10.1093/nar/gkw655.

Love, M.I., Huber, W. and Anders, S. (2014) Moderated estimation of fold change and dispersion for RNA-seq data with DESeq2. *Genome Biology* 15(12), 550. DOI: 10.1186/s13059-014-0550-8.

Martin, M. (2011) Cutadapt removes adapter sequences from high-throughput sequencing reads. *EMBnet Journal* 17(1), 10–12. DOI: 10.14806/ej.17.1.200.

Martin, J., Bruno, V.M., Fang, Z., Meng, X., Blow, M. *et al.* (2010) Rnnotator: an automated de novo transcriptome assembly pipeline from stranded RNA-Seq reads. *BMC Genomics* 11(1), 663. DOI: 10.1186/1471-2164-11-663.

Martin, J.A. and Wang, Z. (2011) Next-generation transcriptome assembly. *Nature Reviews Genetics* 12(10), 671–682. DOI: 10.1038/nrg3068.

McGinnis, S. and Madden, T.L. (2004) BLAST: at the core of a powerful and diverse set of sequence analysis tools. *Nucleic Acids Research* 32(Web Server), W20–W25. DOI: 10.1093/nar/gkh435.

McKenna, A., Hanna, M., Banks, E., Sivachenko, A., Cibulskis, K. *et al.* (2010) The genome analysis toolkit: a MapReduce framework for analyzing next-generation DNA sequencing data. *Genome Research* 20(9), 1297–1303. DOI: 10.1101/gr.107524.110.

McManus, C.J., Duff, M.O., Eipper-Mains, J. and Graveley, B.R. (2010) Global analysis of trans-splicing in *Drosophila*. *Proceedings of the National Academy of Sciences* 107(29), 12975–12979. DOI: 10.1073/pnas.1007586107.

Misof, B., Liu, S., Meusemann, K., Peters, R.S., Donath, A. *et al.* (2014) Phylogenomics resolves the timing and pattern of insect evolution. *Science* 346(6210), 763–767. DOI: 10.1126/science.1257570.

Mitchell, A.L., Attwood, T.K., Babbitt, P.C., Blum, M., Bork, P. *et al.* (2019) InterPro in 2019: improving coverage, classification and access to protein sequence annotations. *Nucleic Acids Research* 47(D1), D351–D360. DOI: 10.1093/nar/gky1100.

Morandin, C., Dhaygude, K., Paviala, J., Trontti, K., Wheat, C. *et al.* (2015) Caste-biases in gene expression are specific to developmental stage in the ant *Formica exsecta*. *Journal of Evolutionary Biology* 28(9), 1705–1718. DOI: 10.1111/jeb.12691.

Nene, V., Wortman, J.R., Lawson, D., Haas, B., Kodira, C. *et al.* (2007) Genome sequence of *Aedes aegypti*, a major arbovirus vector. *Science* 316(5832), 1718–1723. DOI: 10.1126/science.1138878.

Newton, I.L.G., Clark, M.E., Kent, B.N., Bordenstein, S.R., Qu, J. *et al.* (2016) Comparative genomics of two closely related *Wolbachia* with different reproductive effects on hosts. *Genome Biology and Evolution* 8(5), 1526–1542. DOI: 10.1093/gbe/evw096.

Ometto, L., Shoemaker, D., Ross, K.G. and Keller, L. (2010) Evolution of gene expression in fire ants: the effects of developmental stage, caste, and species. *Molecular Biology and Evolution* 28(4), 1381–1392. DOI: 10.1093/molbev/msq322.

Oshlack, A., Robinson, M.D. and Young, M.D. (2010) From RNA-seq reads to differential expression results. *Genome Biology* 11(12), 220. DOI: 10.1186/gb-2010-11-12-220.

Ozsolak, F. and Milos, P.M. (2011) RNA sequencing: advances, challenges and opportunities. *Nature Reviews Genetics* 12(2), 87–98. DOI: 10.1038/nrg2934.

Patel, R.K. and Jain, M. (2012) NGS QC Toolkit: a toolkit for quality control of next generation sequencing data. *PloS ONE* 7(2), e30619. DOI: 10.1371/journal.pone.0030619.

Peng, Y., Leung, H.C.M., Yiu, S.-M., Lv, M.-J., Zhu, X.-G. *et al.* (2013) IDBA-tran: a more robust de novo de Bruijn graph assembler for transcriptomes with uneven expression levels. *Bioinformatics* 29(13), i326–i334. DOI: 10.1093/bioinformatics/btt219.

Pertea, M., Pertea, G.M., Antonescu, C.M., Chang, T.-C., Mendell, J.T. *et al.* (2015) StringTie enables improved reconstruction of a transcriptome from RNA-seq reads. *Nature Biotechnology* 33(3), 290–295. DOI: 10.1038/nbt.3122.

Peters, R.S., Krogmann, L., Mayer, C., Donath, A., Gunkel, S. *et al.* (2017) Evolutionary history of the Hymenoptera. *Current Biology* 27(7), 1013–1018. DOI: 10.1016/j.cub.2017.01.027.

Pimentel, H., Bray, N.L., Puente, S., Melsted, P. and Pachter, L. (2017) Differential analysis of RNA-seq incorporating quantification uncertainty. *Nature Methods* 14(7), 687–690. DOI: 10.1038/nmeth.4324.

Pradeep Kumar, R., John, A., Kumar, P., Dinesh Babu, K.V. and Evans, D.A. (2018) Larvicidal efficacy of Adiantobischrysene from *Adiantum latifolium* against *Oryctes rhinoceros* through disrupting metamorphosis and impeding microbial mediated digestion. *Pest Management Science* 74(8), 1821–1828. DOI: 10.1002/ps.4880.

Price, J., Harrison, M.C., Hammond, R.L., Adams, S., Gutierrez-Marcos, J.F. *et al.* (2018) Alternative splicing associated with phenotypic plasticity in the bumble bee *Bombus terrestris*. *Molecular Ecology* 27(4), 1036–1043. DOI: 10.1111/mec.14495.

Pruitt, K.D., Tatusova, T. and Maglott, D.R. (2005) NCBI Reference Sequence (RefSeq): a curated non-redundant sequence database of genomes, transcripts and proteins. *Nucleic Acids Research* 1, D501–D504.

Qu, W., Gurdziel, K., Pique-Regi, R. and Ruden, D.M. (2017) Identification of splicing quantitative trait loci (sQTL) in Drosophila melanogaster with developmental lead (Pb2+) exposure. *Frontiers in Genetics* 8, 145. DOI: 10.3389/fgene.2017.00145.

Quevillon, E., Silventoinen, V., Pillai, S., Harte, N., Mulder, N. *et al.* (2005) InterProScan: protein domains identifier. *Nucleic Acids Research* 33(Web Server), W116–W120. DOI: 10.1093/nar/gki442.

Ritchie, M.E., Phipson, B., Wu, D., Hu, Y., Law, C.W. *et al.* (2015) limma powers differential expression analyses for RNA-sequencing and microarray studies. *Nucleic Acids Research* 43(7), e47. DOI: 10.1093/nar/gkv007.

Roberts, A. and Pachter, L. (2013) Streaming fragment assignment for real-time analysis of sequencing experiments. *Nature Methods* 10(1), 71–73. DOI: 10.1038/nmeth.2251.

Robertson, G., Schein, J., Chiu, R., Corbett, R., Field, M. *et al.* (2010) De novo assembly and analysis of RNA-seq data. *Nature Methods* 7(11), 909–912. DOI: 10.1038/nmeth.1517.

Robinson, M.D., McCarthy, D.J. and Smyth, G.K. (2010) edgeR: a Bioconductor package for differential expression analysis of digital gene expression data. *Bioinformatics* 26(1), 139–140. DOI: 10.1093/bioinformatics/btp616.

Schmid, R. and Blaxter, M.L. (2008) annot8r: GO, EC and KEGG annotation of EST datasets. *BMC Bioinformatics* 9(1), 180. DOI: 10.1186/1471-2105-9-180.

Schulz, M.H., Zerbino, D.R., Vingron, M. and Birney, E. (2012) Oases: robust de novo RNA-seq assembly across the dynamic range of expression levels. *Bioinformatics* 28(8), 1086–1092. DOI: 10.1093/bioinformatics/bts094.

Shen, S., Park, J.W., Lu, Z.X., Lin, L., Henry, M.D. *et al.* (2014) rMATS: robust and flexible detection of differential alternative splicing from replicate RNA-Seq data. *Proceedings of the National Academy of Sciences* 111(51), E5593–E5601. DOI: 10.1073/pnas.1419161111.

Shi, M., Lin, X.-D., Tian, J.-H., Chen, L.-J., Chen, X. *et al.* (2016) Redefining the invertebrate RNA virosphere. *Nature* 540(7634), 539–543. DOI: 10.1038/nature20167.

Shi, Y. and Jiang, H. (2013) rSeqDiff: detecting differential isoform expression from RNA-Seq data using hierarchical likelihood ratio test. *PLoS ONE* 8(11), e79448. DOI: 10.1371/journal.pone.0079448.

Sim, S.B., Calla, B., Hall, B., DeRego, T. and Geib, S.M. (2015) Reconstructing a comprehensive transcriptome assembly of a white-pupal translocated strain of the pest fruit fly *Bactrocera cucurbitae*. *GigaScience* 4(1), 14. DOI: 10.1186/s13742-015-0053-x.

Simola, D.F., Wissler, L., Donahue, G., Waterhouse, R.M., Helmkampf, M. *et al.* (2013) Social insect genomes exhibit dramatic evolution in gene composition and regulation while preserving regulatory features linked to sociality. *Genome Research* 23(8), 1235–1247. DOI: 10.1101/gr.155408.113.

Standage, D.S., Berens, A.J., Glastad, K.M., Severin, A.J., Brendel, V.P. *et al.* (2016) Genome, transcriptome and methylome sequencing of a primitively eusocial wasp reveal a greatly reduced DNA methylation system in a social insect. *Molecular Ecology* 25(8), 1769–1784. DOI: 10.1111/mec.13578.

Surget-Groba, Y. and Montoya-Burgos, J.I. (2010) Optimization of de novo transcriptome assembly from next-generation sequencing data. *Genome Research* 20(10), 1432–1440. DOI: 10.1101/gr.103846.109.

Tang, S., Lomsadze, A. and Borodovsky, M. (2015) Identification of protein coding regions in RNA transcripts. *Nucleic Acids Research* 43(12), e78. DOI: 10.1093/nar/gkv227.

Tarazona, S., Furió-Tarí, P., Turrà, D., Pietro, A.D., Nueda, M.J. *et al.* (2015) Data quality aware analysis of differential expression in RNA-seq with NOISeq R/Bioc package. *Nucleic Acids Research* 43, e140. DOI: 10.1093/nar/gkv711.

Tatusov, R.L., Fedorova, N.D., Jackson, J.D., Jacobs, A.R., Kiryutin, B. *et al.* (2003) The COG database: an updated version includes eukaryotes. *BMC Bioinformatics* 4(1), 41. DOI: 10.1186/1471-2105-4-41.

Teets, N.M., Peyton, J.T., Colinet, H., Renault, D., Kelley, J.L. *et al.* (2012) Gene expression changes governing extreme dehydration tolerance in an Antarctic insect. *Proceedings of the National Academy of Sciences* 109(50), 20744–20749. DOI: 10.1073/pnas.1218661109.

Thurmond, J., Goodman, J.L., Strelets, V.B., Attrill, H., Gramates, L. S. *et al.* (2019) FlyBase 2.0: the next generation. *Nucleic Acids Research* 47(D1), D759–D765. DOI: 10.1093/nar/gky1003.

Tian, J., Wang, Z., Mei, S., Yang, N., Yang, Y. *et al.* (2018) CancerSplicingQTL: a database for genome-wide identification of splicing QTLs in human cancer. *Nucleic Acids Research* 47(D1), D909–D916. DOI: 10.1093/nar/gky954.

Trapnell, C., Hendrickson, D.G., Sauvageau, M., Goff, L., Rinn, J.L. *et al.* (2013) Differential analysis of gene regulation at transcript resolution with RNA-seq. *Nature Biotechnology* 31(1), 46–53. DOI: 10.1038/nbt.2450.

UniProt (2019) UniProt: a worldwide hub of protein knowledge. *Nucleic Acids Research* 47(D1), D506–D515. DOI: 10.1093/nar/gky1049.

Trapnell, C., Roberts, A., Goff, L., Pertea, G., Kim, D. *et al.* (2012) Differential gene and transcript expression analysis of RNA-seq experiments with TopHat and Cufflinks. *Nature Protocols* 7(3), 562–578. DOI: 10.1038/nprot.2012.016.

Vaquero-Garcia, J., Barrera, A., Gazzara, M.R., González-Vallinas, J., Lahens, N.F. *et al.* (2016) A new view of transcriptome complexity and regulation through the lens of local splicing variations. *elife* 5, e11752. DOI: 10.7554/eLife.11752.

Wagner, G.P., Kin, K. and Lynch, V.J. (2012) Measurement of mRNA abundance using RNA-seq data: RPKM measure is inconsistent among samples. *Theory in Biosciences* 131(4), 281–285. DOI: 10.1007/s12064-012-0162-3.

Wang, L., Park, H.J., Dasari, S., Wang, S., Kocher, J.-P. *et al.* (2013a) CPAT: Coding-Potential Assessment Tool using an alignment-free logistic regression model. *Nucleic Acids Research* 41(6), e74. DOI: 10.1093/nar/gkt006.

Wang, W., Qin, Z., Feng, Z., Wang, X. and Zhang, X. (2013b) Identifying differentially spliced genes from two groups of RNA-seq samples. *Gene* 518(1), 164–170. DOI: 10.1016/j.gene.2012.11.045.

Wang, B., Tseng, E., Regulski, M., Clark, T.A., Hon, T. *et al.* (2016) Unveiling the complexity of the maize transcriptome by single-molecule long-read sequencing. *Nature communications* 7(1), 11708. DOI: 10.1038/ncomms11708.

Wang, L., Wang, S. and Li, W. (2012) RSeQC: quality control of RNA-seq experiments. *Bioinformatics* 28(16), 2184–2185. DOI: 10.1093/bioinformatics/bts356.

Wipfler, B., Letsch, H., Frandsen, P.B., Kapli, P., Mayer, C. *et al.* (2019) Evolutionary history of Polyneoptera and its implications for our understanding of early winged insects. *Proceedings of the National Academy of Sciences* 116(8), 3024–3029. DOI: 10.1073/pnas.1817794116.

Wu, D.C., Yao, J., Ho, K.S., Lambowitz, A.M. and Wilke, C.O. (2018) Limitations of alignment-free tools in total RNA-seq quantification. *BMC Genomics* 19(1), 510. DOI: 10.1186/s12864-018-4869-5.

Xia, Q., Zhou, Z., Lu, C., Cheng, D., Dai, F. *et al.* (2004) A draft sequence for the genome of the domesticated silkworm (*Bombyx mori*). *Science* 306, 1937–1940.

Xie, Y., Wu, G., Tang, J., Luo, R., Patterson, J. *et al.* (2014) SOAPdenovo-Trans: de novo transcriptome assembly with short RNA-Seq reads. *Bioinformatics* 30(12), 1660–1666. DOI: 10.1093/bioinformatics/btu077.

Xiong, H.Y., Alipanahi, B., Lee, L.J., Bretschneider, H., Merico, D. *et al.* (2015) The human splicing code reveals new insights into the genetic determinants of disease. *Science* 347(6218), 1254806. DOI: 10.1126/science.1254806.

Yin, C., Liu, Y., Liu, J., Xiao, H., Huang, S. *et al.* (2014) ChiloDB: a genomic and transcriptome database for an important rice insect pest *Chilo suppressalis*. *Database* 2014, bau065. DOI: 10.1093/database/bau065.

Yin, C., Shen, G., Guo, D., Wang, S., Ma, X. *et al.* (2016) InsectBase: a resource for insect genomes and transcriptomes. *Nucleic Acids Research* 44(D1), D801–D807. DOI: 10.1093/nar/gkv1204.

Yu, Y., Zhou, H., Kong, Y., Pan, B., Chen, L. *et al.* (2016) The landscape of A-to-I RNA editome is shaped by both positive and purifying selection. *PLoS Genetics* 12(7), e1006191. DOI: 10.1371/journal.pgen.1006191.

Zeng, V. and Extavour, C.G. (2012) ASGARD: an open-access database of annotated transcriptomes for emerging model arthropod species. *Database* 2012, bas048. DOI: 10.1093/database/bas048.

Zhang, H.-M., Chen, H., Liu, W., Liu, H., Gong, J. *et al.* (2011) AnimalTFDB: a comprehensive animal transcription factor database. *Nucleic Acids Research* 40(D1), D144–D149. DOI: 10.1093/nar/gkr965.

Zhang, Y.-Z., Shi, M. and Holmes, E.C. (2018) Using metagenomics to characterize an expanding virosphere. *Cell* 172(6), 1168–1172. DOI: 10.1016/j.cell.2018.02.043.

Zhang, C., Zhang, B., Lin, L.-L. and Zhao, S. (2017) Evaluation and comparison of computational tools for RNA-seq isoform quantification. *BMC Genomics* 18(1), 583. DOI: 10.1186/s12864-017-4002-1.

Zhou, C., Liu, S., Song, W., Luo, S., Meng, G. *et al.* (2018) Characterization of viral RNA splicing using whole-transcriptome datasets from host species. *Scientific Reports* 8(1), 3273. DOI: 10.1038/s41598-018-21190-7.

3 Transcriptomics in Pest Management Research

Antonino Malacrinò*

Department of Evolution, Ecology and Organismal Biology, The Ohio State University, Columbus, Ohio, USA

3.1 Introduction

While farmers have had to deal with pests since the development of agriculture, the intensification and simplification of agroecosystems has increased pest damage to crops. In response, farmers have developed physical, biological, and chemical control measures to limit crop damage and economic losses. The development of agrochemicals (especially fertilizers and pesticides) has boosted crop yields around the world by satisfying crops' nutritional requirements and protecting them from biological threats (e.g. weeds, pathogens, animal pests). However, it has been estimated that pests still cause the loss of 37% of potential crop yield in the USA (Pimentel, 2005). The negative impact of pesticides on health and the environment are increasingly apparent (Chagnon *et al.*, 2015). In Europe for example, their impact on human health produces an annual cost of €78 million (Fantke *et al.*, 2012). Other issues include environmental contamination (including effects on non-target species and water resources), residues in food, and the onset of resistance (Carvalho, 2006). Thus, one of the biggest challenges for agricultural science is to develop a pest control protocol that can lower pest population densities under the damage threshold without harming the environment and public health. This challenge is of high interest in the scientific community, so researchers across the globe are trying to solve it using the most advanced techniques currently available.

High-throughput sequencing (HTS) techniques represent a keystone to unveiling genomes, transcriptomes, and proteomes at different levels of resolution, from single cells to entire communities (Zhang *et al.*, 2010). As these technologies have changed rapidly in recent years, the way genomes and transcriptomes (which previously relied on cloning thousands of fragments

*antonino.malacrino@gmail.com, malacrino.1@osu.edu

and sequencing them with low-throughput approaches) are analyzed has also changed. Initially incurring extremely high costs and time-spans for data analysis, massive amounts of data can be now obtained within a few days or even hours and analyzed almost in real time at a fraction of the cost compared with the past (Reuter *et al.*, 2015).

Knowledge of an organism's transcriptome and the transcriptomic response to biotic and abiotic factors is extremely powerful. First, immediate feedback on the genes that an organism expresses or about the pathways activated as a response to specific environmental variables can be obtained. The transcriptome can also be used to dig for finer information and use it to develop innovative control strategies. Recently, this approach has been utilized to exploit the post-transcriptional regulatory mechanism known as RNA interference (RNAi) in order to knockdown the expression of specific genes, resulting in knocking out the targeted insect species (Gordon and Waterhouse, 2007; Swevers and Smagghe, 2012; Burand and Hunter, 2013; Laudani *et al.*, 2017). Transcriptomics is making a significant contribution in this field of research by allowing the screening of potential new targets for RNAi (Wang *et al.*, 2011).

Plant–herbivore interaction represents another field with wide-ranging applications. Although plants have developed an array of defenses against primary consumers, herbivory is widespread; it has been estimated that every plant species represents the food source of at least one insect species (Schoonhoven *et al.*, 2005). The study of insect transcriptomics has revealed how insects respond to plant defenses and which strategies they use to thrive (Vogel *et al.*, 2014; Thorpe *et al.*, 2016; Schweizer *et al.*, 2017). A comprehensive understanding of these mechanisms is fundamental to developing new pest management techniques targeted either to enhance plant defenses or harm pest strategies to overcome those defenses.

Another aspect of insect physiology that has implications in pest management is the occurrence of diapause. Some insect species can exploit this mechanism to overcome unfavorable environmental conditions; for others, this step is part of their ontogeny. Transcriptomics has been used in these cases to study the physiology of diapause comprehensively and identify the regulatory pathways behind it (Tu *et al.*, 2015; Koštál *et al.*, 2017), which can then be targeted for pest management purposes.

Therefore, it is understandable that analyzing the entire set of genes expressed by an organism can allow a comprehensive analysis of its biology, physiology, and ecology. With regard to plant protection, transcriptomics represents one of the most powerful tools currently available to fully investigate insect pests and find new and environmentally friendly ways to control them. The focus of this chapter is to summarize how transcriptomics has boosted research in pest science, what lessons it teaches, and possible directions for future study. The goal is to review the literature and highlight the biggest areas for future research: (i) transcriptome characterization of insect pests, invasive species, and biocontrol agents; (ii) understanding the evolution and mechanisms behind pesticide resistance; and (iii) dissecting insect–microorganism interactions.

3.2 Insect Pests, Invasive Species and Biocontrol Agents

As outlined above, in-depth analysis of the transcriptomes of insect pests can reveal important information about their biology, physiology and behavior. Previous studies have focused on this field and explored the gene expression of insect populations under field and laboratory conditions. Significant effort has focused on pests of interest in plant protection, and, currently, the transcriptomes of several species are available for further investigation. The scope of this section is to give an overview of the breadth of research in pest science using transcriptomics.

Tephritid flies represent a serious threat to different crops, and many studies focused on their transcriptome to better understand their biology, with the aim of finding potential targets for their control (Shen *et al.*, 2011; Sim *et al.*, 2015; Wang *et al.*, 2016; Pavlidi *et al.*, 2017; Kumaran *et al.*, 2018). Additional research focused on other polyphagous species (*Spodoptera* spp. and *Helicoverpa* spp.) when feeding on different hosts and characterized their specific responses to host variation (Roy *et al.*, 2016; Cheng *et al.*, 2017; Pearce *et al.*, 2017). The Colorado potato beetle *Leptinotarsa decemlineata* Say has also been investigated under transcriptomics, revealing a wide set of genes coding for detoxifying enzymes and duplications in the RNAi pathway (Kumar *et al.*, 2014; Schoville *et al.*, 2018). The different pathways that insect pests use to interact with their host, and the strategies they develop to overcome host defenses, represent a potential target for their control.

Phenotypic plasticity is also a phenomenon that has been investigated through transcriptomics. An example is *Schistocerca gregaria* (Forskål), whose phenotype can change from a cryptic solitarious phase to a swarming gregarious phase. When comparing the transcriptome between the two phenotypes, Badisco *et al.* (2011) revealed changes in the expression of genes coding for central nervous system development, sensory perception, stress response, and fundamental cellular processes. In another study, Zhai *et al.* (2013) compared high- and low-fecundity populations of *Nilaparvata lugens* (Stål), a serious rice pest in Asia, using a coupled transcriptomics/proteomics approach. The analyses resulted in a list of candidate genes likely regulating fecundity in *N. lugens*, including a glutamine synthetase gene, which was knocked down using RNAi and resulted in a subsequent reduction in fecundity. Qin *et al.* (2017) studied the population dynamics of the grasshopper *Chorthippus albonemus* Zheng & Tu in reaction to grazing by other herbivores in the same area and found a response in the insect's transcriptome, especially in ankyrin repeat domain-containing proteins and heat shock proteins. Forest pests have also been investigated using transcriptomics. The transcriptome of *Dendroctonus ponderosae* Hopkins has been studied by Keeling *et al.* (2012) and Nadeau *et al.* (2017), who compared male and female individuals in order to characterize their pheromone biosynthesis pathways.

The promise of transcriptomics has also attracted research on invasive insect species. *Rhynchophorus ferrugineus* (Olivier) is responsible for the destruction of commercial and ornamental palm species in invaded habitats, and its transcriptomes have been characterized in its antennae (Antony *et al.*, 2016) and during early development (Yin *et al.*, 2015). Other studies targeted the emerald ash borer *Agrilus planipennis* Fairmaire, which is devastating ash forests in North America; they described its whole body transcriptome and also investigated tissue-specific gene expression (Mittapalli *et al.*, 2010; Mamidala *et al.*, 2013; Duan *et al.*, 2015). In another study, Tarver *et al.* (2016) characterized the transcriptome of *Aethina tumida* Murray, a pest of honey bees (Palmeri *et al.*, 2015), across life stages, sexes, and different tissues.

Several other pests have also been characterized and their transcriptomes are available for further investigation: *Apolygus lucorum* (Meyer-Dür) (Cao *et al.*, 2016), *Anthonomus grandis* Boheman (Salvador *et al.*, 2014), *Plutella xylostella* (L.) (Jouraku *et al.*, 2013), *Sitobion avenae* F. and *Acyrthosiphon pisum* Harris (Wang *et al.*, 2014), *Rhopalosiphum padi* (L.) (Duan *et al.*, 2017), *Lutzomyia longipalpis* (Lutz and Neiva) (McCarthy *et al.*, 2013), *Migdolus fryanus* Westwood (Nakayama *et al.*, 2017), *Musca domestica* L. (Liu *et al.*, 2012), *Sogatella furcifera* (Horváth) (Li *et al.*, 2016), and *Heliothis virescens* (F.) (Perera *et al.*, 2015). The transcriptomes of several major cowpea pests have been characterized by Agunbiade *et al.* (2013). Several further species of insect pests have been analyzed with the aim of characterizing their resistance to insecticides (see below). Transcriptomics has also proved to be useful in studying the venom of parasitoid wasps (Yan *et al.*, 2016; Teng *et al.*, 2017), bringing new information on their evolution and mode of action, likely impacting future biocontrol techniques (Nakasu *et al.*, 2014; Moreau *et al.*, 2015).

The 1KITE project (1K Insect Transcriptome Evolution) (http://www.1kite.org) is also a noteworthy source of information; it details the transcriptome sequencing of more than 1200 diverse insect species. When the 1KITE dataset was compared with the European and Mediterranean Plant Protection Organization (EPPO) Global database, a collection of pests of interest to plant protection (https://gd.eppo.int/), 85 species were identified as insect pests and 36 as biocontrol agents, making 1KITE a powerful source of information to develop the next generation of pest management programs.

3.3 Pesticide Resistance

The use of synthetic chemicals to control pests (pesticides) had striking effects and the practice quickly became widely used globally. The evolution of resistance to pesticides, one of the drawbacks of their indiscriminate use, can lead to the complete failure of pest management strategies, negatively impacting crop yield and the environment with little or no benefit. The Arthropod Pesticide Resistance Database (www.pesticideresistance.org), a project aimed at collecting information regarding the onset of resistance to pesticides worldwide, includes 16,862 cases of 604 arthropod species (February, 2019). These data are even more alarming considering that during 2008 the number of cases reported was 7747 (Onstad, 2014), and 51 species have been added to the database in the past 9 years (i.e. reported as resistant to a pesticide for the first time).

Unfortunately, the evolution of this phenomenon and its genetic and genomic aspects are not yet fully understood.

Resistance to pesticides represents a fundamental example of human-influenced natural selection, so studies of the underlying mechanisms are of both academic and applied interest. Our understanding of the molecular basis of insecticide resistance and how it translates into the spread of resistant alleles in nature is still in its infancy and is largely derived from intra- and inter-specific comparisons. The major factors governing resistance to pesticides are known, as are the major loci involved in resistance to different classes of pesticides (ffrench-Constant *et al.*, 2004; ffrench-Constant, 2013). Natural populations can evolve resistance to pesticides in different ways, including increasing or decreasing the expression of key genes (Feyereisen *et al.*, 2015). Resistance to insecticides can evolve through changes in the target sites (e.g. receptors or other proteins), which decrease the sensitivity to the pesticide, as well as through changes in expression or efficiency of detoxifying enzymes (e.g. cytochrome P450, esterases, glutathione S-transferases) (ffrench-Constant, 2014). These responses could be due to mutations affecting the whole gene, copy number variation, or *cis* or *trans* regulation. Earlier studies suggested that the development of resistance impacts gene expression in a more complicated fashion than previously thought (Pedra *et al.*, 2004; Silva *et al.*, 2012; Qiu *et al.*, 2013) and may involve effects on energy metabolism, detoxifying enzymes, proteins of extracellular transport, peptidases and cuticular proteins, and hundreds of other genes.

HTS technologies produced a revolution in this field, allowing in-depth comparison of the transcriptomes of resistant/non-resistant populations in both field and laboratory conditions. This aspect was investigated in several species: *Aedes albopictus* Skuse (Grigoraki *et al.*, 2015), *Anopheles arabiensis* Patton (Jones *et al.*, 2012), *Aphis gossypii* Glover (Pan *et al.*, 2015a, 2015b), *Apis mellifera* L. (Aufauvre *et al.*, 2014), *Bemisia tabaci* (Gennadius) (Yang *et al.*, 2013a), *Cimex lectularius* L. (Mamidala *et al.*, 2012), *Culex pipiens* L. (Lv *et al.*, 2016), *Musca domestica* L. (Li *et al.*, 2013), *Plutella xylostella* (L.) (He *et al.*, 2012), *Sogatella furcifera* (Horváth) (Yang *et al.*, 2016), and *Trialeurodes vaporariorum* Westwood (Karatolos *et al.*, 2011). Although these studies spanned different insect species and pesticides (e.g. Deltamethrin, DDT, Chlorpyrifos, etc.), they consistently reported the overexpression of several cytochrome P450 genes, genes for different esterases and transferases, and some ATP-binding cassette genes (ABC). Other studies only analyzed the genes activated after exposure to selected insecticides, without comparing them with a resistant population. According to these studies, both *B. tabaci* (Yang *et al.*, 2013b) and *L. decemlineata* (Clements *et al.*, 2018) increased their expression of P450, transferase, and esterase genes when exposed to insecticides. Another category of studies searched across the transcriptome of different pests and identified genes (mainly P450s, transferases, esterases) likely to play a role in pesticide detoxification: *Bactrocera dorsalis* (Hendel) (Hsu *et al.*, 2012), *B. tabaci* (Wang *et al.*, 2010; Xie *et al.*, 2012; Ye *et al.*, 2014), *C. lectularius* (Adelman *et al.*, 2011), *Cnaphalocrocis medinalis* (Guenée) (Yu *et al.*, 2015), *Cryptolaemus montrouzieri* Mulsant (Zhang *et al.*, 2012), *Lymantria dispar* (L.) (Cao *et al.*, 2015),

Mythimna separata Walker (Liu *et al.*, 2016), and *Nilaparvata lugens* (Stål) (Lao *et al.*, 2015).

In summary, several studies used transcriptomics to describe the effect of pesticides on gene expression, and of gene expression on pesticide resistance. However, detailed information on the evolution of resistance and the contribution of secondary loci is lacking. One way to address this challenge is to allow populations of model organisms to adapt to controlled laboratory environments, an approach known as experimental evolution. Researchers are now able to evolve populations to new conditions and re-sequence their genomes and transcriptomes (evolve and re-sequence approach, E&R), enabling real-time monitoring of molecular evolution on a genome-wide scale (Long *et al.*, 2015). The characterization of regulatory modules in gene expression is pivotal in understanding the evolution of resistance and could be used together with E&R to thoroughly study the development of resistance to insecticides. Re-sequencing the genome and transcriptome can provide real-time information on the trajectories of the loci that contribute to the adaptation. This approach has been used in *Drosophila* to uncover genomic details related to body size variation (Turner *et al.*, 2011), adaptation to hypoxia (Zhou *et al.*, 2011), and increased survival after virus infection (Martins *et al.*, 2014). It has also been extensively applied to other organisms (Long *et al.*, 2015).

3.4 Insect–Microorganism Interaction

It is widely acknowledged that microorganisms are the main drivers of several fundamental physical, chemical, and biological phenomena (Prosser, 2015). With their ubiquitous distribution, they can structure and modulate several biotic interactions through their interactions with other organisms (Whitman *et al.*, 1998; Prosser, 2015). Indeed, the study of microbial ecology is spreading throughout the scientific community, starting with human microbiology and expanding to other research areas from marine ecology to food and insect sciences (Huber *et al.*, 2007; Douglas, 2009; Ercolini, 2013).

Recently, the microbial communities associated with insects have attracted increasing interest, mainly because of their ecological and economic importance. Microorganisms possess metabolic properties that are often absent in insects. Therefore, they can act as 'brokers', for example enabling insects to overcome plant defenses that they could not on their own (Dillon and Dillon, 2004). The interaction between microbes and insects is not new to the scientific community, since it is widely acknowledged that bacteria can help insects thrive on nutritionally marginal diets (Buchner, 1965). Evolution led insects to exploit microorganisms in order to adapt to nutrient-limited environments by feeding directly on microorganisms or relegating them to pre-digest refractory diets (Engel and Moran, 2013). This exploitation is particularly clear in insect herbivores, since plant tissues contain a wide range of indigestible and toxic compounds, and thus would not be ideal foods for insects. Therefore, insects evolved a series of strategies to cope with these issues (Goldman-Huertas *et al.*, 2015), including associating with bacterial and fungal symbionts to exploit different hosts (Frago *et al.*, 2012; Brady and White, 2013). The interaction between insects and microorganisms can extend to different levels of the trophic ladder and can also comprise indirect interactions mediated by plants (Biere and Bennett, 2013). Moreover, microorganisms can influence insects' responses to plant defenses, provide protection from natural enemies, and influence the reproductive system (Ferrari and Vavre, 2011). These examples are just a few illustrations of the plasticity of the insect–microorganism relationships. Sometimes this association is so strict that experimental exchanges of gut microbial communities in insect herbivores lead to a decrease in performance on the host plant (Hosokawa *et al.*, 2007).

The study of insect–microorganism interactions was significantly advanced by the diffusion of 'omics' techniques in microbial ecology, which allowed the response of microbes to environmental factors to be studied in greater detail (Abdelfattah *et al.*, 2018; Malacrinò, 2018). The main way these technologies have been exploited in microbial ecology is to perform meta'omics studies, which aim to identify microbial communities, their genes along with their expression, and the metabolic pathways and functions in environmental and uncultured samples (Segata *et al.*, 2014; Peršoh, 2015). A meta-transcriptomic approach can not only be used for taxonomical reconstruction (provided there is enough information on transcriptomes), but also tells which genes the community is expressing in that sample (Shi *et al.*, 2010). So, in addition to knowing who is inhabiting the sample, information on which biological processes are active is also available.

In addition to symbionts, transcriptomics can help identify microbial pathogens. A meta-transcriptomic approach was used by Cox-Foster *et al.* (2007) to investigate the causal agent of colony collapse disorder in honey bees; their findings suggested a correlation between the disease and the detection of Israeli acute paralysis virus, although a metagenomics study discounted this relationship (Honeybee Genome Sequencing Consortium, 2006). Other authors using metatranscriptomics have identified numerous bee pathogens, some only recently described (Tozkar *et al.*, 2015). A similar approach allowed the identification of candidate pathogens responsible for the decline of the invasive yellow crazy ant *Anoplolepis gracilipes* F. Smith in Australia by providing pivotal information that can be integrated into pest management programs (Cooling *et al.*, 2017). Transcriptomics has also been used to study the response of insect pests to pathogenesis processes by entomopathogenic fungi and viruses (Zhang *et al.*, 2015; Chu *et al.*, 2016; Xing *et al.*, 2017).

3.5 Conclusions

Modern molecular technologies and bioinformatic tools may help us find alternative ways to protect crops, forests, and ecosystems while preserving human health and the environment. These developments are particularly relevant since current climatic and social challenges are triggering radical changes in the way plant protection is approached. Scientists are assisting the increased efforts for sustainable agriculture by exploiting genomic technologies to retrieve pivotal information to design effective controls for insect pests. Technologies and protocols are constantly improving, procedure automation is becoming increasingly common, and modern bioinformatic and machine-learning workflows are helping the transition from conventional approaches to these new techniques. These advancements are accomplished by decreasing both the time-to-results and the range of skills required to process samples correctly and retrieve meaningful information. Therefore, now is the time to climb through the looking-glass, search beyond the laboratory bench, and push our research efforts into real-world situations.

References

Abdelfattah, A., Malacrinò, A., Wisniewski, M., Cacciola, S.O. and Schena, L. (2018) Metabarcoding: a powerful tool to investigate microbial communities and shape future plant protection strategies. *Biological Control* 120, 1–10. DOI: 10.1016/j.biocontrol.2017.07.009.

Adelman, Z.N., Kilcullen, K.A., Koganemaru, R., Anderson, M.A.E., Anderson, T.D. *et al.* (2011) Deep sequencing of pyrethroid-resistant bed bugs reveals multiple mechanisms of resistance within a single population. *PLoS ONE* 6(10), e26228. DOI: 10.1371/journal.pone.0026228.

Agunbiade, T.A., Sun, W., Coates, B.S., Djouaka, R., Tamò, M. *et al.* (2013) Development of reference transcriptomes for the major field insect pests of cowpea: a toolbox for insect pest management approaches in West Africa. *PLoS ONE* 8(11), e79929. DOI: 10.1371/journal.pone.0079929.

Antony, B., Soffan, A., Jakše, J., Abdelazim, M.M., Aldosari, S.A. *et al.* (2016) Identification of the genes involved in odorant reception and detection in the palm weevil *Rhynchophorus ferrugineus*, an important quarantine pest, by antennal transcriptome analysis. *BMC Genomics* 17(1), 69. DOI: 10.1186/s12864-016-2362-6.

Aufauvre, J., Misme-Aucouturier, B., Viguès, B., Texier, C., Delbac, F. *et al.* (2014) Transcriptome analyses of the honeybee response to *Nosema ceranae* and insecticides. *PLoS ONE* 9(3), e91686. DOI: 10.1371/journal.pone.0091686.

Badisco, L., Ott, S.R., Rogers, S.M., Matheson, T., Knapen, D. *et al.* (2011) Microarray-based transcriptomic analysis of differences between long-term gregarious and solitarious desert locusts. *PLoS ONE* 6(11), e28110. DOI: 10.1371/journal.pone.0028110.

Biere, A. and Bennett, A.E. (2013) Three-way interactions between plants, microbes and insects. *Functional Ecology* 27(3), 567–573. DOI: 10.1111/1365-2435.12100.

Brady, C.M. and White, J.A. (2013) Cowpea aphid (*Aphis craccivora*) associated with different host plants has different facultative endosymbionts. *Ecological Entomology* 38(4), 433–437. DOI: 10.1111/een.12020.

Buchner, P. (1965) *Endosymbiosis of Animals with Plant Microorganisms.* Interscience Publishers, New York.

Burand, J.P. and Hunter, W.B. (2013) RNAi: future in insect management. *Journal of Invertebrate Pathology* 112, S68–S74. DOI: 10.1016/j.jip.2012.07.012.

Cao, D.-P., Liu, Y., Wei, J.-J., Li, J.-H. and Wang, G.-R. (2016) A *de novo* transcriptomic analysis to reveal functional genes in *Apolygus lucorum*. *Insect Science* 23(1), 2–14. DOI: 10.1111/1744-7917.12188.

Cao, C., Sun, L., Wen, R., Shang, Q., Ma, L. *et al.* (2015) Characterization of the transcriptome of the Asian gypsy moth *Lymantria dispar* identifies numerous transcripts associated with insecticide resistance. *Pesticide Biochemistry and Physiology* 119, 54–61. DOI: 10.1016/j.pestbp.2015.02.005.

Carvalho, F.P. (2006) Agriculture, pesticides, food security and food safety. *Environmental Science & Policy* 9(7–8), 685–692. DOI: 10.1016/j.envsci.2006.08.002.

Chagnon, M., Kreutzweiser, D., Mitchell, E.A.D., Morrissey, C.A., Noome, D.A. *et al.* (2015) Risks of large-scale use of systemic insecticides to ecosystem functioning and services. *Environmental Science and Pollution Research* 22(1), 119–134. DOI: 10.1007/s11356-014-3277-x.

Cheng, T., Wu, J., Wu, Y., Chilukuri, R.V., Huang, L. *et al.* (2017) Genomic adaptation to polyphagy and insecticides in a major East Asian noctuid pest. *Nature Ecology & Evolution* 1(11), 1747–1756. DOI: 10.1038/s41559-017-0314-4.

Chu, Z.-J., Wang, Y.-J., Ying, S.-H., Wang, X.-W. and Feng, M.-G. (2016) Genome-wide host–pathogen interaction unveiled by transcriptomic response of diamondback moth to fungal infection. *PLoS ONE* 11(4), e0152908. DOI: 10.1371/journal.pone.0152908.

Clements, J., Sanchez-Sedillo, B., Bradfield, C.A. and Groves, R.L. (2018) Transcriptomic analysis reveals similarities in genetic activation of detoxification mechanisms resulting from imidacloprid and chlorothalonil exposure. *PLoS ONE* 13(10), e0205881. DOI: 10.1371/journal.pone.0205881.

Cooling, M., Gruber, M.A.M., Hoffmann, B.D., Sébastien, A. and Lester, P.J. (2017) A metatranscriptomic survey of the invasive yellow crazy ant, *Anoplolepis gracilipes*, identifies several potential viral and bacterial pathogens and mutualists. *Insectes Sociaux* 64(2), 197–207. DOI: 10.1007/s00040-016-0531-x.

Cox-Foster, D.L., Conlan, S., Holmes, E.C., Palacios, G., Evans, J.D. *et al.* (2007) A metagenomic survey of microbes in honey bee colony collapse disorder. *Science* 318(5848), 283–728. DOI: 10.1126/science.1146498.

Dillon, R.J. and Dillon, V.M. (2004) The gut bacteria of insects : nonpathogenic interactions. *Annual Review of Entomology* 49(1), 71–92.

Douglas, A.E. (2009) The microbial dimension in insect nutritional ecology. *Functional Ecology* 23(1), 38–47. DOI: 10.1111/j.1365-2435.2008.01442.x.

Duan, J., Ladd, T., Doucet, D., Cusson, M., VanFrankenhuyzen, K. *et al.* (2015) Transcriptome analysis of the emerald ash borer (EAB), *Agrilus planipennis*: de novo assembly, functional annotation and comparative analysis. *PLoS ONE* 10(8), e0134824. DOI: 10.1371/journal.pone.0134824.

Duan, X., Wang, K., Su, S., Tian, R., Li, Y. *et al.* (2017) De novo transcriptome analysis and microsatellite marker development for population genetic study of a serious insect pest, *Rhopalosiphum padi* (L.) (Hemiptera: Aphididae). *PLoS ONE* 12(2), e0172513. DOI: 10.1371/journal.pone.0172513.

Engel, P. and Moran, N.A. (2013) The gut microbiota of insects – diversity in structure and function. *FEMS Microbiology Reviews* 37(5), 699–735. DOI: 10.1111/1574-6976.12025.

Ercolini, D. (2013) High-throughput sequencing and metagenomics: moving forward in the culture-independent analysis of food microbial ecology. *Applied and Environmental Microbiology* 79(10), 3148–3155. DOI: 10.1128/AEM.00256-13.

Fantke, P., Friedrich, R. and Jolliet, O. (2012) Health impact and damage cost assessment of pesticides in Europe. *Environment International* 49, 9–17. DOI: 10.1016/j.envint.2012.08.001.

Ferrari, J. and Vavre, F. (2011) Bacterial symbionts in insects or the story of communities affecting communities. *Philosophical Transactions of the Royal Society of London. Series B, Biological Sciences* 366(1569), 1389–1400.

Feyereisen, R., Dermauw, W. and Van Leeuwen, T. (2015) Genotype to phenotype, the molecular and physiological dimensions of resistance in arthropods. *Pesticide Biochemistry and Physiology* 121, 61–77. DOI: 10.1016/j.pestbp.2015.01.004.

ffrench-Constant, R.H. (2013) The molecular genetics of insecticide resistance. *Genetics* 194(4), 807–815. DOI: 10.1534/genetics.112.141895.

ffrench-Constant, R.H. (2014) Insecticide resistance comes of age. *Genome Biology* 15(2), 106. DOI: 10.1186/gb4162.

ffrench-Constant, R.H., Daborn, P.J. and Goff, G.L. (2004) The genetics and genomics of insecticide resistance. *Trends in Genetics* 20(3), 163–170. DOI: 10.1016/j.tig.2004.01.003.

Frago, E., Dicke, M. and Godfray, H.C.J. (2012) Insect symbionts as hidden players in insect–plant interactions. *Trends in Ecology & Evolution* 27(12), 705–711. DOI: 10.1016/j.tree.2012.08.013.

Goldman-Huertas, B., Mitchell, R.F., Lapoint, R.T., Faucher, C.P., Hildebrand, J.G. *et al.* (2015) Evolution of herbivory in *Drosophilidae* linked to loss of behaviors, antennal responses, odorant receptors, and ancestral diet. *Proceedings of the National Academy of Sciences of the United States of America* 112(10), 3026–3031. DOI: 10.1073/pnas.1424656112.

Gordon, K.H.J. and Waterhouse, P.M. (2007) RNAi for insect-proof plants. *Nature Biotechnology* 25(11), 1231–1232. DOI: 10.1038/nbt1107-1231.

Grigoraki, L., Lagnel, J., Kioulos, I., Kampouraki, A., Morou, E. *et al.* (2015) Transcriptome profiling and genetic study reveal amplified carboxylesterase genes implicated in temephos resistance, in the Asian tiger mosquito *Aedes albopictus*. *PLoS Neglected Tropical Diseases* 9(5), e0003771. DOI: 10.1371/journal.pntd.0003771.

He, W., You, M., Vasseur, L., Yang, G., Xie, M. *et al.* (2012) Developmental and insecticide-resistant insights from the de novo assembled transcriptome of the diamondback moth, *Plutella xylostella*. *Genomics* 99(3), 169–177. DOI: 10.1016/j.ygeno.2011.12.009.

Honeybee Genome Sequencing Consortium (2006) Insights into social insects from the genome of the honeybee, *Apis mellifera*. *Nature* 443(7114), 931–949. DOI: 10.1038/nature05260.

Hosokawa, T., Kikuchi, Y., Shimada, M. and Fukatsu, T. (2007) Obligate symbiont involved in pest status of host insect. *Proceedings of the Royal Society B: Biological Sciences* 274(1621), 1979–1984. DOI: 10.1098/rspb.2007.0620.

Hsu, J.-C., Chien, T.-Y., Hu, C.-C., Chen, M.-J.M., Wu, W.-J. *et al.*and (2012) Discovery of genes related to insecticide resistance in *Bactrocera dorsalis* by functional genomic analysis of a de novo assembled transcriptome. *PLoS ONE* 7(8), e40950. DOI: 10.1371/journal.pone.0040950.

Huber, J.A., Mark Welch, D.B., Morrison, H.G., Huse, S.M., Neal, P.R. *et al.* (2007) Microbial population structures in the deep marine biosphere. *Science* 318(5847), 97–100. DOI: 10.1126/science.1146689.

Jones, C.M., Toé, H.K., Sanou, A., Namountougou, M., Hughes, A. *et al.* (2012) Additional selection for insecticide resistance in urban malaria vectors: DDT resistance in *Anopheles arabiensis* from Bobo-Dioulasso, Burkina Faso. *PLoS ONE* 7(9), e45995. DOI: 10.1371/journal.pone.0045995.

Jouraku, A., Yamamoto, K., Kuwazaki, S., Urio, M., Suetsugu, Y. *et al.* (2013) KONAGAbase: a genomic and transcriptomic database for the diamondback moth, *Plutella xylostella*. *BMC Genomics* 14(1), 464. DOI: 10.1186/1471-2164-14-464.

Karatolos, N., Pauchet, Y., Wilkinson, P., Chauhan, R., Denholm, I. *et al.* (2011) Pyrosequencing the transcriptome of the greenhouse whitefly, *Trialeurodes vaporariorum* reveals multiple transcripts encoding insecticide targets and detoxifying enzymes. *BMC Genomics* 12(1), 56. DOI: 10.1186/1471-2164-12-56.

Keeling, C.I., Henderson, H., Li, M., Yuen, M., Clark, E.L. *et al.* (2012) Transcriptome and full-length cDNA resources for the mountain pine beetle, *Dendroctonus ponderosae* Hopkins, a major insect pest of pine forests. *Insect Biochemistry and Molecular Biology* 42(8), 525–536. DOI: 10.1016/j.ibmb.2012.03.010.

Košťál, V., Štětina, T., Poupardin, R., Korbelová, J. and Bruce, A.W. (2017) Conceptual framework of the eco-physiological phases of insect diapause development justified by transcriptomic profiling. *Proceedings of the National Academy of Sciences of the United States of America* 114(32), 8532–8537. DOI: 10.1073/pnas.1707281114.

Kumar, A., Congiu, L., Lindström, L., Piiroinen, S., Vidotto, M. *et al.* (2014) Sequencing, de novo assembly and annotation of the Colorado potato beetle, *Leptinotarsa decemlineata*, transcriptome. *PLoS ONE* 9(1), e86012. DOI: 10.1371/journal.pone.0086012.

Kumaran, N., van der Burg, C.A., Qin, Y., Cameron, S.L., Clarke, A.R. *et al.* (2018) Plant-mediated female transcriptomic changes post-mating in a tephritid fruit fly, *Bactrocera tryoni*. *Genome Biology and Evolution* 10(1), 94–107. DOI: 10.1093/gbe/evx257.

Lao, S.-H., Huang, X.-H., Huang, H.-J., Liu, C.-W., Zhang, C.-X. *et al.* (2015) Genomic and transcriptomic insights into the cytochrome P450 monooxygenase gene repertoire in the rice pest brown planthopper, *Nilaparvata lugens*. *Genomics* 106(5), 301–309. DOI: 10.1016/j.ygeno.2015.07.010.

Laudani, F., Strano, C.P., Edwards, M.G., Malacrinò, A., Campolo, O. *et al.* (2017) RNAi-mediated gene silencing in Rhynchophorus ferrugineus (Oliver) (Coleoptera: Curculionidae). *Open Life Sciences* 12(1), 214–222. DOI: 10.1515/biol-2017-0025.

Li, Z., An, X.-K., Liu, Y.-D. and Hou, M.-L. (2016) Transcriptomic and expression analysis of the salivary glands in white-backed planthoppers, *Sogatella furcifera*. *PLoS ONE* 11(7), e0159393. DOI: 10.1371/journal.pone.0159393.

Li, M., Reid, W.R., Zhang, L., Scott, J.G., Gao, X. *et al.* (2013) A whole transcriptomal linkage analysis of gene co-regulation in insecticide resistant house flies, *Musca domestica*. *BMC Genomics* 14(1), 803. DOI: 10.1186/1471-2164-14-803.

Liu, Y., Qi, M., Chi, Y. and Wuriyanghan, H. (2016) De novo assembly of the transcriptome for oriental armyworm *Mythimna separata* (Lepidoptera: Noctuidae) and analysis on insecticide resistance-related genes. *Journal of Insect Science* 16(1), 92. DOI: 10.1093/jisesa/iew079.

Liu, F., Tang, T., Sun, L. and Jose Priya, T.A. (2012) Transcriptomic analysis of the housefly (*Musca domestica*) larva using massively parallel pyrosequencing. *Molecular Biology Reports* 39(2), 1927–1934. DOI: 10.1007/s11033-011-0939-3.

Long, A., Liti, G., Luptak, A. and Tenaillon, O. (2015) Elucidating the molecular architecture of adaptation via evolve and resequence experiments. *Nature Reviews Genetics* 16(10), 567–582. DOI: 10.1038/nrg3937.

Lv, Y., Wang, W., Hong, S., Lei, Z., Fang, F. *et al.* (2016) Comparative transcriptome analyses of deltamethrin-susceptible and -resistant *Culex pipiens* pallens by RNA-seq. *Molecular Genetics and Genomics* 291(1), 309–321. DOI: 10.1007/s00438-015-1109-4.

Malacrinò, A. (2018) Meta-Omics tools in the world of insect–microorganism interactions. *Biology* 7(4), 50. DOI: 10.3390/biology7040050.

Mamidala, P., Wijeratne, A.J., Wijeratne, S., Kornacker, K., Sudhamalla, B. *et al.* (2012) RNA-Seq and molecular docking reveal multi-level pesticide resistance in the bed bug. *BMC Genomics* 13(1), 6. DOI: 10.1186/1471-2164-13-6.

Mamidala, P., Wijeratne, A.J., Wijeratne, S., Poland, T., Qazi, S.S. *et al.* (2013) Identification of odor-processing genes in the emerald ash borer, *Agrilus planipennis*. *PLoS ONE* 8(2), e56555. DOI: 10.1371/journal.pone.0056555.

Martins, N.E., Faria, V.G., Nolte, V., Schlötterer, C., Teixeira, L. *et al.* (2014) Host adaptation to viruses relies on few genes with different cross-resistance properties. *Proceedings of the National Academy of Sciences of the United States of America* 111(16), 5938–5943. DOI: 10.1073/pnas.1400378111.

McCarthy, C.B., Santini, M.S., Pimenta, P.F.P. and Diambra, L.A. (2013) First comparative transcriptomic analysis of wild adult male and female *Lutzomyia longipalpis*, vector of visceral leishmaniasis. *PLoS ONE* 8(3), e58645. DOI: 10.1371/journal.pone.0058645.

Mittapalli, O., Bai, X., Mamidala, P., Rajarapu, S.P., Bonello, P. *et al.* (2010) Tissue-specific transcriptomics of the exotic invasive insect pest emerald ash borer (*Agrilus planipennis*). *PLoS ONE* 5(10), e13708. DOI: 10.1371/journal.pone.0013708.

Moreau, S. Asgari, S., Moreau, S.J.M. and Asgari, S. (2015) Venom proteins from parasitoid wasps and their biological functions. *Toxins* 7(7), 2385–2412. DOI: 10.3390/toxins7072385.

Nadeau, J.A., Petereit, J., Tillett, R.L., Jung, K., Fotoohi, M. *et al.* (2017) Comparative transcriptomics of mountain pine beetle pheromone-biosynthetic tissues and functional analysis of CYP6DE3. *BMC Genomics* 18(1), 311. DOI: 10.1186/s12864-017-3696-4.

Nakasu, E.Y.T., Williamson, S.M., Edwards, M.G., Fitches, E.C., Gatehouse, J.A. *et al.* (2014) Novel biopesticide based on a spider venom peptide shows no adverse effects on honeybees. *Proceedings of the Royal Society B: Biological Sciences* 281(1787), 20140619. DOI: 10.1098/rspb.2014.0619.

Nakayama, D.G., Santos Júnior, C.D., Kishi, L.T., Pedezzi, R., Santiago, A.C. *et al.* (2017) A transcriptomic survey of *Migdolus fryanus* (sugarcane rhizome borer) larvae. *PLoS ONE* 12(3), e0173059. DOI: 10.1371/journal.pone.0173059.

Onstad, D.W. (2014) Major issues in insect resistance management. In: Onstad, D.W. (ed.) *Insect Resistance Management*, 2nd edn. Academic Press, Wilmington, Delaware, pp. 1–23.

Palmeri, V., Scirtò, G., Malacrinò, A., Laudani, F. and Campolo, O. (2015) A scientific note on a new pest for European honeybees: first report of small hive beetle *Aethina tumida*, (Coleoptera: Nitidulidae) in Italy. *Apidologie* 46(4), 527–529. DOI: 10.1007/s13592-014-0343-9.

Pan, Y., Peng, T., Gao, X., Zhang, L., Yang, C. *et al.* (2015a) Transcriptomic comparison of thiamethoxam-resistance adaptation in resistant and susceptible strains of *Aphis gossypii* Glover. *Comparative Biochemistry and Physiology Part D: Genomics and Proteomics* 13, 10–15. DOI: 10.1016/j.cbd.2014.11.001.

Pan, Y., Yang, C., Gao, X., Peng, T., Bi, R. *et al.* (2015b) Spirotetramat resistance adaption analysis of *Aphis gossypii* Glover by transcriptomic survey. *Pesticide Biochemistry and Physiology* 124, 73–80. DOI: 10.1016/j.pestbp.2015.04.007.

Pavlidi, N., Gioti, A., Wybouw, N., Dermauw, W., Ben-Yosef, M. *et al.* (2017) Transcriptomic responses of the olive fruit fly *Bactrocera oleae* and its symbiont Candidatus *Erwinia dacicola* to olive feeding. *Scientific Reports* 7(1), 42633. DOI: 10.1038/srep42633.

Pearce, S.L., Clarke, D.F., East, P.D., Elfekih, S., Gordon, K.H.J. *et al.* (2017) Genomic innovations, transcriptional plasticity and gene loss underlying the evolution and divergence of two highly polyphagous and invasive *Helicoverpa* pest species. *BMC Biology* 15(1), 63. DOI: 10.1186/s12915-017-0402-6.

Pedra, J.H.F., McIntyre, L.M., Scharf, M.E. and Pittendrigh, B.R. (2004) Genome-wide transcription profile of field- and laboratory-selected dichlorodiphenyltrichloroethane (DDT)-resistant *Drosophila*. *Proceedings of the National Academy of Sciences* 101(18), 7034–7039. DOI: 10.1073/pnas.0400580101.

Perera, O.P., Shelby, K.S., Popham, H.J.R., Gould, F., Adang, M.J. *et al.* (2015) Generation of a transcriptome in a model lepidopteran pest, *Heliothis virescens*, using multiple sequencing strategies for profiling midgut gene expression. *PLoS ONE* 10(6), e0128563. DOI: 10.1371/journal.pone.0128563.

Peršoh, D. (2015) Plant-associated fungal communities in the light of meta'omics. *Fungal Diversity* 75(1), 1–25. DOI: 10.1007/s13225-015-0334-9.

Pimentel, D. (2005) Environmental and economic costs of the application of pesticides primarily in the United States. *Environment, Development and Sustainability* 7(2), 229–252. DOI: 10.1007/s10668-005-7314-2.

Prosser, J.I. (2015) Dispersing misconceptions and identifying opportunities for the use of 'omics' in soil microbial ecology. *Nature Reviews Microbiology* 13(7), 439–446. DOI: 10.1038/nrmicro3468.

Qin, X., Ma, J., Huang, X., Kallenbach, R.L., Lock, T.R. *et al.* (2017) Population dynamics and transcriptomic responses of *Chorthippus albonemus* (Orthoptera: Acrididae) to herbivore grazing intensity. *Frontiers in Ecology and Evolution* 5, 136. DOI: 10.3389/fevo.2017.00136.

Qiu, X., Sun, W., McDonnell, C.M., Li-Byarlay, H., Steele, L.D. *et al.* (2013) Genome-wide analysis of genes associated with moderate and high DDT resistance in *Drosophila melanogaster*. *Pest Management Science* 69(8), 930–937. DOI: 10.1002/ps.3454.

Reuter, J.A., Spacek, D.V. and Snyder, M.P. (2015) High-Throughput Sequencing Technologies. *Molecular Cell* 58(4), 586–597. DOI: 10.1016/j.molcel.2015.05.004.

Roy, A., Walker, W.B., Vogel, H., Chattington, S., Larsson, M.C. *et al.* (2016) Diet dependent metabolic responses in three generalist insect herbivores *Spodoptera* spp. *Insect Biochemistry and Molecular Biology* 71, 91–105. DOI: 10.1016/j.ibmb.2016.02.006.

Salvador, R., Príncipi, D., Berretta, M., Fernández, P., Paniego, N. *et al.* (2014) Transcriptomic survey of the midgut of *Anthonomus grandis* (Coleoptera: Curculionidae). *Journal of Insect Science* 14(1), 219. DOI: 10.1093/jisesa/ieu081.

Schoonhoven, L.M., Loon, J.J.Avan. and Dicke, M. (2005) *Insect–Plant Biology*. Oxford University Press, New York.

Schoville, S.D., Chen, Y.H., Andersson, M.N., Benoit, J.B., Bhandari, A. *et al.* (2018) A model species for agricultural pest genomics: the genome of the Colorado potato beetle, *Leptinotarsa decemlineata* (Coleoptera: Chrysomelidae). *Scientific Reports* 8(1), 1931. DOI: 10.1038/s41598-018-20154-1.

Schweizer, F., Heidel-Fischer, H., Vogel, H. and Reymond, P. (2017) *Arabidopsis* glucosinolates trigger a contrasting transcriptomic response in a generalist and a specialist herbivore. *Insect Biochemistry and Molecular Biology* 85, 21–31. DOI: 10.1016/j.ibmb.2017.04.004.

Segata, N., Boernigen, D., Tickle, T.L., Morgan, X.C., Garrett, W.S. *et al.* (2014) Computational meta'omics for microbial community studies. *Molecular Systems Biology* 9(1), 666. DOI: 10.1038/msb.2013.22.

Shen, G.-M., Dou, W., Niu, J.-Z., Jiang, H.-B., Yang, W.-J. *et al.* (2011) Transcriptome analysis of the oriental fruit fly (*Bactrocera dorsalis*). *PLoS ONE* 6(12), e29127. DOI: 10.1371/journal.pone.0029127.

Shi, W., Syrenne, R., Sun, J.-Z. and Yuan, J.S. (2010) Molecular approaches to study the insect gut symbiotic microbiota at the 'omics' age. *Insect Science* 17(3), 199–219. DOI: 10.1111/j.1744-7917.2010.01340.x.

Silva, A.X., Jander, G., Samaniego, H., Ramsey, J.S. and Figueroa, C.C. (2012) Insecticide resistance mechanisms in the green peach aphid *Myzus persicae* (Hemiptera: Aphididae) I: a transcriptomic survey. *PLoS ONE* 7(6), e36366. DOI: 10.1371/journal.pone.0036366.

Sim, S.B., Calla, B., Hall, B., DeRego, T. and Geib, S.M. (2015) Reconstructing a comprehensive transcriptome assembly of a white-pupal translocated strain of the pest fruit fly *Bactrocera cucurbitae*. *GigaScience* 4(1), 14. DOI: 10.1186/s13742-015-0053-x.

Swevers, L. and Smagghe, G. (2012) Use of RNAi for control of insect crop pests. In: Smagghe, G. and Diaz, I. (eds) *Arthropod–Plant Interactions*. Springer Netherlands, Dordrecht, Netherlands, pp. 177–197.

Tarver, M.R., Huang, Q., de Guzman, L., Rinderer, T., Holloway, B. *et al.* (2016) Transcriptomic and functional resources for the small hive beetle *Aethina tumida*, a worldwide parasite of honey bees. *Genomics Data* 9, 97–99. DOI: 10.1016/j.gdata.2016.06.003.

Teng, Z.-W., Xiong, S.-J., Xu, G., Gan, S.-Y., Chen, X. *et al.* (2017) Protein discovery: combined transcriptomic and proteomic analyses of venom from the endoparasitoid *Cotesia chilonis* (Hymenoptera: Braconidae). *Toxins* 9(4), 135. DOI: 10.3390/toxins9040135.

Thorpe, P., Cock, P.J.A. and Bos, J. (2016) Comparative transcriptomics and proteomics of three different aphid species identifies core and diverse effector sets. *BMC Genomics* 17(1), 172. DOI: 10.1186/s12864-016-2496-6.

Tozkar, C.Ö., Kence, M., Kence, A., Huang, Q. and Evans, J.D. (2015) Metatranscriptomic analyses of honey bee colonies. *Frontiers in Genetics* 6, 100. DOI: 10.3389/fgene.2015.00100.

Tu, X., Wang, J., Hao, K., Whitman, D.W., Fan, Y. *et al.* (2015) Transcriptomic and proteomic analysis of pre-diapause and non-diapause eggs of migratory locust, *Locusta migratoria* L. (Orthoptera: Acridoidea). *Scientific Reports* 5(1), 11402. DOI: 10.1038/srep11402.

Turner, T.L., Stewart, A.D., Fields, A.T., Rice, W.R. and Tarone, A.M. (2011) Population-based resequencing of experimentally evolved populations reveals the genetic basis of body size variation in *Drosophila melanogaster*. *PLoS Genetics* 7(3), e1001336. DOI: 10.1371/journal.pgen.1001336.

Vogel, H., Musser, R.O. and de la Paz Celorio-Mancera, M. (2014) Transcriptome responses in herbivorous insects towards host plant and toxin feeding. In: Voelckel, C. and Jander, G. (eds) *Annual Plant Reviews: Insect–Plant Interactions*. John Wiley & Sons, Chichester, UK, pp. 197–233.

Wang, D., Liu, Q., Jones, H.D., Bruce, T. and Xia, L. (2014) Comparative transcriptomic analyses revealed divergences of two agriculturally important aphid species. *BMC Genomics* 15(1), 1023. DOI: 10.1186/1471-2164-15-1023.

Wang, X.-W., Luan, J.-B., Li, J.-M., Bao, Y.-Y., Zhang, C.-X. *et al.* (2010) De novo characterization of a whitefly transcriptome and analysis of its gene expression during development. *BMC Genomics* 11(1), 400. DOI: 10.1186/1471-2164-11-400.

Wang, J., Xiong, K.-C. and Liu, Y.-H. (2016) De novo transcriptome analysis of Chinese citrus fly, *Bactrocera minax* (Diptera: Tephritidae), by high-throughput Illumina sequencing. *PLoS ONE* 11(6), e0157656. DOI: 10.1371/journal.pone.0157656.

Wang, Y., Zhang, H., Li, H. and Miao, X. (2011) Second-generation sequencing supply an effective way to screen RNAi targets in large scale for potential application in pest insect control. *PLoS ONE* 6(4), e18644. DOI: 10.1371/journal.pone.0018644.

Whitman, W.B., Coleman, D.C. and Wiebe, W.J. (1998) Prokaryotes: the unseen majority. *Proceedings of the National Academy of Sciences of the United States of America* 95(12), 6578–6583. DOI: 10.1073/pnas.95.12.6578.

Xie, W., Meng, Q., Wu, Q., Wang, S., Yang, X. *et al.* (2012) Pyrosequencing the *Bemisia tabaci* transcriptome reveals a highly diverse bacterial community and a robust system for insecticide resistance. *PLoS ONE* 7(4), e35181. DOI: 10.1371/journal.pone.0035181.

Xing, L., Yuan, C., Wang, M., Lin, Z., Shen, B. *et al.* (2017) Dynamics of the interaction between cotton bollworm *Helicoverpa armigera* and Nucleopolyhedrovirus as revealed by integrated transcriptomic and proteomic analyses. *Molecular & Cellular Proteomics* 16(6), 1009–1028. DOI: 10.1074/mcp.M116.062547.

Yan, Z., Fang, Q., Wang, L., Liu, J., Zhu, Y. *et al.* (2016) Insights into the venom composition and evolution of an endoparasitoid wasp by combining proteomic and transcriptomic analyses. *Scientific Reports* 6(1), 19604. DOI: 10.1038/srep19604.

Yang, N., Xie, W., Jones, C.M., Bass, C., Jiao, X. *et al.* (2013a) Transcriptome profiling of the whitefly *Bemisia tabaci* reveals stage-specific gene expression signatures for thiamethoxam resistance. *Insect Molecular Biology* 22(5), 485–496. DOI: 10.1111/imb.12038.

Yang, N., Xie, W., Yang, X., Wang, S., Wu, Q. *et al.* (2013b) Transcriptomic and proteomic responses of sweetpotato whitefly, *Bemisia tabaci*, to thiamethoxam. *PLoS ONE* 8(5), e61820. DOI: 10.1371/journal.pone.0061820.

Yang, Y., Zhang, Y., Yang, B., Fang, J. and Liu, Z. (2016) Transcriptomic responses to different doses of cycloxaprid involved in detoxification and stress response in the whitebacked planthopper, *Sogatella furcifera*. *Entomologia Experimentalis et Applicata* 158(3), 248–257. DOI: 10.1111/eea.12406.

Ye, X.-D., Su, Y.-L., Zhao, Q.-Y., Xia, W.-Q., Liu, S.-S. *et al.* (2014) Transcriptomic analyses reveal the adaptive features and biological differences of guts from two invasive whitefly species. *BMC Genomics* 15(1), 370. DOI: 10.1186/1471-2164-15-370.

Yin, A., Pan, L., Zhang, X., Wang, L., Yin, Y. *et al.* (2015) Transcriptomic study of the red palm weevil *Rhynchophorus ferrugineus* embryogenesis. *Insect Science* 22(1), 65–82. DOI: 10.1111/1744-7917.12092.

Yu, H.-Z., Wen, D.-F., Wang, W.-L., Geng, L., Zhang, Y. *et al.* (2015) Identification of genes putatively involved in chitin metabolism and insecticide detoxification in the rice leaf folder (*Cnaphalocrocis medinalis*) larvae through transcriptomic analysis. *International Journal of Molecular Sciences* 16(9), 21873–21896. DOI: 10.3390/ijms160921873.

Zhai, Y., Zhang, J., Sun, Z., Dong, X., He, Y. *et al.* (2013) Proteomic and transcriptomic analyses of fecundity in the brown planthopper *Nilaparvata lugens* (Stål). *Journal of Proteome Research* 12(11), 5199–5212. DOI: 10.1021/pr400561c.

Zhang, W., Li, F. and Nie, L. (2010) Integrating multiple 'omics' analysis for microbial biology: application and methodologies. *Microbiology* 156(2), 287–301. DOI: 10.1099/mic.0.034793-0.

Zhang, W., Chen, J., Keyhani, N.O., Zhang, Z., Li, S. *et al.* (2015) Comparative transcriptomic analysis of immune responses of the migratory locust, *Locusta migratoria*, to challenge by the fungal insect pathogen, *Metarhizium acridum*. *BMC Genomics* 16(1), 867. DOI: 10.1186/s12864-015-2089-9.

Zhang, Y., Jiang, R., Wu, H., Liu, P., Xie, J. *et al.* (2012) Next-generation sequencing-based transcriptome analysis of *Cryptolaemus montrouzieri* under insecticide stress reveals resistance-relevant genes in ladybirds. *Genomics* 100(1), 35–41. DOI: 10.1016/j.ygeno.2012.05.002.

Zhou, D., Udpa, N., Gersten, M., Visk, D.W., Bashir, A. *et al.* (2011) Experimental selection of hypoxia-tolerant *Drosophila melanogaster*. *Proceedings of the National Academy of Sciences of the United States of America* 108(6), 2349–2354. DOI: 10.1073/pnas.1010643108.

4 Aphid Transcriptomics – Past, Present and Future

Sampurna Sattar,* and Gary A. Thompson*

College of Agricultural Sciences, The Pennsylvania State University, University Park, PA, USA

4.1 Introduction

Rapid development of cost-effective new sequencing technologies with improved sensitivity have led to the advent of the transcriptomic era in organismal research. Hundreds of organisms and their specialized tissues have been subjected to transcriptomic analysis to understand the functional genetics associated with environmental stress, disease state, and specific developmental stages. Transcriptomics complements whole genome sequencing analysis and is an effective method to denote functions to unannotated genes. The earliest transcriptome studies involved preparation of complementary DNA (cDNA) libraries and Sanger sequencing to generate expression sequence tags (ESTs) for a subset of the transcripts. Real-time PCR (q-PCR) assays and microarrays were used to determine the expression of specific transcripts that would ultimately contribute to improved understanding of gene functions. Both these methods are labor intensive, compared with the new generation of sequencing technologies, and proved to be unsuitable for large-scale experiments involving whole transcriptome sequencing and genome annotation.

Advancement of high-throughput, next-generation sequencing (NGS) technologies opened a wide range of possibilities in transcriptome research. NGS transcriptome sequencing that can record 10^9 transcript sequences is now routinely used to sequence transcriptomes from both model and non-model organisms (Hornett and Wheat, 2012; Lowe *et al.*, 2017; Chen *et al.*, 2018). The emergence of RNA-seq (RNA-sequencing), which employs NGS technology, allows transcript quantification and enables mapping to a reference genome or transcriptome (Wang *et al.*, 2009). This technology, although initially used for transcript profiling of humans and model organisms such as yeast, *Arabidopsis*, and mice, is increasingly used for non-model organisms

*Corresponding author: sus56@psu.edu; gat10@psu.edu

with little or no prior genome or transcriptome information (Cloonan *et al.*, 2008; Lister *et al.*, 2008; Nagalakshmi *et al.*, 2008; Mortazavi *et al.*, 2008; Hornett and Wheat, 2012). Unlike microarrays, which require pre-existing genome or transcriptome information for the preparation of probes, RNA-seq technology directly sequences transcripts and provides information on splice variants, and is especially useful if limited RNA is available (Ozsolak and Milos, 2011; Lowe *et al.*, 2017). In brief, RNA-seq library preparation starts with enrichment of poly-A RNA that is converted to cDNA segments. The cDNA strands are ligated to adapter sequences at either one or both ends and subjected to sequencing using NGS platforms to yield short reads of 300–400 bases that are either single-end or paired-end reads (Wang *et al.*, 2009). These sequences are then processed by stringent computational analysis to be assembled into a transcriptome, or they can be mapped to an existing genome or reference transcriptome. Transcript abundance can be compared between two states, such as healthy and diseased tissue samples, to measure the effect of environmental stressors or to measure the effect of chemical application on organisms. The extensive *in silico* sequence analysis that is required for data interpretation from these experiments was considered to be a bottleneck, requiring interdisciplinary collaborations between computer scientists and biologists. However, this field of research has received considerable attention in the past decade and great strides have been made in computational analysis and management of NGS data (Nussinov, 2015). By overcoming these challenges, transcriptomics has rapidly evolved in the past decade and has allowed research to advance in non-model organisms, including several insect species.

4.1.1 Brief overview of insect genomic and transcriptomic research

Insects are the most diverse and abundant group of invertebrate hexapods, belonging to the phylum Arthropoda that evolved around 480 million years ago in the Ordovician era (Misof *et al.*, 2014). In recent years a better understanding of the phylogenetic and evolutionary relationships of different insect species has emerged as a direct consequence of a high-throughput sequencing of insect genomes and transcriptomes. The model insect *Drosophila melanogaster* was among the first to have its genome sequenced completly. Within the next decade several more insect species, including mosquito and silkworm, were added to this growing list (Boerjan *et al.*, 2012). Insect transcriptomic research was pursued in parallel with advances in genomic studies. In rapid succession, microarrays and EST libraries from *Drosophila*, mosquito, honey bees, flour beetles, and silkworm were reported and submitted to public databases (Oppenheim *et al.*, 2014). As NGS platforms developed and became easily available, collaborative efforts began to emerge among entomologists, leading to the formation of the i5K international consortia with an immediate goal to sequence and annotate 5000 arthropod genomes in 5 years (http://i5k.github.io/about). In addition, the 1KITE (1K

Insect Transcriptome Evolution) project was established with the aim to study insect transcriptomes representing all insect orders (http://1kite.org).

4.1.2 Hemipteran insects: specialized mouthparts for unique feeding habit

Hemipteran insects are economically important agricultural, veterinary, and medical pests, due to their unique feeding habits, rapid reproductive rates, and their ability to transmit pathogens (Koch *et al.*, 2016). Hemipterans or true bugs are found across the globe and are presently divided into four sub-orders: Auchenorrhyncha (represented by cicadas, leafhoppers, treehoppers, planthoppers), Coleorrhyncha (represented by moss bugs), Heteroptera (represented by assassin bugs, bed bugs, shield bugs), and Sternorrhyncha (represented by aphids and whiteflies) (Carver *et al.*, 1991; Dolling, 1991; Song *et al.*, 2012; Szwedo, 2018). Most Hemiptera feed on plant sap and cause considerable damage to agricultural crops either through direct feeding or indirectly by transmission of plant pathogens. A common feature among all Hemiptera is their unique feeding habit using specialized sucking mouth-parts that are modified to form stylets, which can be used for piercing and sucking plant sap or animal body fluids. The ultrastructure of the stylet shows that it consists of a bundle made from mandibular and maxillary stylets that are interlocked in a groove (Goodchild, 1966). Internal to the maxillary stylet are the salivary and food canals; one is for saliva movement into the host tissue and the other for the uptake of plant sap by the insect. A salivary pump and cibarial pump drive saliva and ingest plant sap, respectively, and both pumps are under the control of dilator muscles in the head (Panfilio and Angelini, 2018). The stylet can penetrate host tissue to reach feeding sites in vasculature such as the xylem or phloem or the parenchyma, depending on the species (Tjallingii and Hogen Esch, 1993). These insects also produce two kinds of saliva: watery saliva and gelling saliva. The watery saliva is introduced into the host tissue and often contains hydrolyzing enzymes and metabolites to suppress plant defenses, whereas the gelling saliva forms a sheath around the stylet to seal the puncture site (Miles, 1972). Feeding behavior associated with stylet movement, sap ingestion and salivation within the plant tissues can be documented with an electro-penetration graph (EPG) (Tjallingii, 1985).

4.1.3 Hemipteran transcriptomics

The unique set of adaptations seen in hemipteran insects has raised questions not only about their evolutionary biology, but also from the application standpoint of establishing well-defined integrated pest management (IPM) strategies. So far, several phenotypically diverse Hemiptera species have been chosen for developing genomic resources. These studies have mostly used a comparative approach between species, developmental stages, tissues types, sexual and asexual morphs, and host interactions to understand the molecular basis for insect physiology and relate it to phenotype and insect behaviors. Most phloem-feeding hemipteran insects also harbor symbiotic bacteria that are maternally inherited (Buchner, 1965; Douglas, 2006). For example, transcriptomic studies in pea aphids and whiteflies focusing on their nutritional uptake have cued in on the importance of symbiotic bacteria for insect survival, fecundity, and virus transmission (Burke and Moran, 2011; Sseruwagi *et al.*, 2018). Other studies on the whitefly gut transcriptome have provided crucial information on xenobiotic degradation for insect survival (Ye *et al.*, 2014). Emerging reports of resistance to chemical pesticides have also prompted transcriptome-wide analyses of insect resistance and detoxification genes that will inform future IPM strategies (Karatolos *et al.*, 2011; Silva *et al.*, 2012). In addition to gut-specific transcriptome studies, analysis of salivary transcriptomes has enhanced our understanding of the roles of salivary proteins in insect survival and the silencing of host defense mechanisms (DeLay *et al.*, 2012; Ji *et al.*, 2013). Another common theme of transcriptome-wide analyses is the study of chemosensory and odorant-binding genes from head and antennal transcriptomes, revealing information about proteins that are essential for olfactory behaviors such as searching for food, mating, and oviposition (Gu *et al.*, 2013; Zhou *et al.*, 2014; Wang *et al.*, 2017; Bian *et al.*, 2018). Transcriptome-wide analysis of immune response genes from two different species of brown planthoppers has resulted in a better understanding of the innate immune response during virus transmission (Bao *et al.*, 2013; Zhao *et al.*, 2016). A similar study in whiteflies found changes in the insects' transcriptome in response to tomato yellow leaf curl virus infection (Hasegawa *et al.*, 2018). Transcriptome analysis of the developmental stages of whiteflies and the plant bug *Adelphocoris suturalis* has led to an improved understanding of their basic biology (Wang *et al.*, 2010; Tian *et al.*, 2015).

Several invasive species that have been introduced into diverse geographical regions and pose an imminent economic threat to agriculture have caught the attention of the research community. Efforts to build genomic resources for the invasive species are contributing to a better understanding of their basic biology and directing control efforts. For example, the full transcriptome sequence was recently completed for the brown marmorated stink bug, an invasive species introduced to the north-eastern USA (Ioannidis *et al.*, 2014).

4.2 Aphid Transcriptomics

Aphids are phloem-feeding Hemiptera belonging to the superfamily Aphidoidea. Agriculturally important aphid species that mostly feed on herbaceous plants are placed in the subfamily Aphididae, whereas the closely related woolly adelgids infesting coniferous forest species are placed in the subfamily Adelgidae (Blackman and Eastop, 2007). Aphids, like other members of Hemiptera, are sap-sucking insects with specialized mouthparts; however, their reproductive plasticity makes them unique within the taxa. The alternation of sexual reproduction and asexual parthenogenesis is an exclusive feature of aphids that has been instrumental in their success as herbivores. Telescoping of generations is another reproductive capability of aphids that expedites clonal generation (Kindlmann and Dixon, 1989). As a result of this strategy, in the summer months an aphid colony can rapidly expand in the field, leading to extensive pest pressure and severe crop damage. In these situations, application of chemical pesticides is the preferred mode of control; however, many aphid species have developed resistance to chemical pesticides (Field and Devonshire, 1998). Despite their importance as agricultural pests, aphids lack appreciable genomic resources as compared with the widely studied *Drosophila*.

4.2.1 Genomic resources for studying aphid biology

The need for genomic resources to study aphid biology led to the creation of AphidBase as a repository of aphid genomic and transcriptomic data and to serve as a functional annotation tool (Gauthier *et al.*, 2007). A concerted effort by the research community selected the pea aphid *Acyrthosiphon pisum* as a model hemipteran. It was the first hemipteran with a fully sequenced genome that was subsequently annotated (Tagu *et al.*, 2008; The International Aphid Genomics Consortium, 2010). Several seminal studies of pea aphid revealed new information on reproductive plasticity, salivary gland protein gene expression, seasonal photoperiodism, evolutionary divergence, and symbiont association (Cortés *et al.*, 2008; Legeai *et al.*, 2010; Shigenobu *et al.*, 2010; Burke and Moran, 2011; Carolan *et al.*, 2011; Le Trionnaire *et al.*, 2012). While research flourished in pea aphid, genomic resources for other economically important aphid species were lacking. In the past decade, however, the whole genomes of the Russian wheat aphid *Diuraphis noxia* and soybean aphid *Aphis glycines* and draft genomes of the green peach aphid *Myzus persicae* and the melon and cotton aphid *Aphis gossypii* have been published (BIPAA, n.d.; Nicholson *et al.*, 2015; Wenger *et al.*, 2017; Quan *et al.*, 2019). The genomes of other economically important aphid species such as the greenbug *Schizaphis graminum* and yellow sugarcane aphid *Sipha flava* are currently being sequenced and assembled (http://i5k.github.io/arthropod_genomes_at_ncbi). In addition to genomes, robust collections of ESTs from *M. persicae*, *A. gossypii*, the bird cherry oat aphid *Rhopalosiphum padi*, and the brown citrus aphid *Toxoptera citricida* have been made available in public databases (Hunter *et al.*, 2003; Tagu *et al.*, 2004; Figueroa *et al.*, 2007; Ramsey *et al.*, 2007).

4.2.2 Pea aphid – paradigm for aphid transcriptomics

Early transcriptomic studies of *Acyrthosiphon pisum* cDNA libraries created a catalogue of 50,599 full-length cDNA clones, resulting in 97,828 good-quality ESTs representing 9000 aphid genes (Cortés *et al.*, 2008; Shigenobu *et al.*, 2010). These ESTs deposited in AphidBase proved useful for the initial round of genome annotation of the 464 Mb pea aphid draft genome. The collection of sequenced transcripts in AphidBase continued to expand, and their functional annotation also improved through validation experiments. In one of the earliest transcriptomic studies, the salivary protein C002 was identified from cDNA libraries prepared from pea aphid salivary glands (Mutti *et al.*, 2008). RNAi-mediated knockdown studies in the pea aphid, and subsequently in other species such as green peach aphid and green bug, have confirmed that this salivary protein is essential for aphid feeding, fecundity, and survival on host plants (Mutti *et al.*, 2006, 2008; Pitino *et al.*, 2011; Zhang *et al.*, 2015).

The molecular mechanisms regulating aphid developmental plasticity (polyphenism) have long been of interest to evolutionary biologists, and with the advent of genomic tools the blueprint for a reproductive mode switch is now well understood. Large-scale transcript profiling of aphids reared during different photoperiods identified neuro-endrocrine signaling genes regulating this process (Le Trionnaire *et al.*, 2012). Earlier studies on reproductive mode switching conducted in controlled conditions could only identify 10% of the significantly regulated transcripts (Le Trionnaire *et al.*, 2007, 2009). Only two cuticular and three signaling genes were significantly regulated in different reproductive morphs in response to seasonal photoperiodism (Le Trionnaire *et al.*, 2007; Gallot *et al.*, 2010). In outdoor conditions, the effect of seasonal photoperiod and temporal changes induced increased differences in the transcriptome of developing aphids (Le Trionnaire *et al.*, 2012).

Other types of polyphenisms triggered by environmental cues include wing dimorphism, which is often observed in response to crowding when female, asexual, viviparous aphids develop embryos with wings, allowing for the next generation to immediately disperse to new food source (Brisson, 2010). In an initial study, cDNA microarray was used to compare the transcript profiles in winged and wingless populations in both males and females. Differential expression of transcripts regulating energy production was observed between winged and wingless males. Overall differences in the transcripts were more pronounced for winged and wingless morphs in males, compared with winged and wingless females (Brisson *et al.*, 2007). Differential gene expression patterns were studied between winged and wingless morphs to understand the genetic network-regulating wing development in aphids (Brisson, 2010). Comparative expression profiling of a subset of the developmental genes – apterous (*ap*), decapentaplegic (*dpp*), Distalless (*Dll*), engrailed (*en*), hedgehog (*hh*), homothorax (*hth*), Ultrabithorax (*Ubx*), and wingless (*wg*) – was performed in winged and wingless morphs and the expression of *ap1* was significantly higher in first and second instar winged morphs (Brisson, 2010).

A complete molecular signature associated with aphid crowding was revealed from a genome-wide study of the maternal transcriptome. In this study, significant transcriptional reprogramming of genes regulating odorant binding, hormonal activity, and neurotransmitter transport was reported in response to high density populations (Vellichirammal *et al.*, 2016). Body color polymorphism in aphids is another ecologically important trait that can be triggered by environmental changes or may be genetically determined (Tsuchida *et al.*, 2010). The color is due to pigments present in aphid cuticles and internal tissues, hemolymph and bacteriocytes (Tsuchida, 2016). A recent study comparing the transcriptome profile of red and green morphs of pea aphid revealed that only a small subset of transcripts were differentially regulated between the two morphs. Gene-encoding carotenoid dehydrogenase and energy-related gene V-type H^+ ATPase were significantly up-regulated in the red morph (Zhang *et al.*, 2018).

A more comprehensive approach to investigating the fundamental development processes such as axis formation, neurogenesis, eye development, wing development, and germline specification was undertaken by Shigenobu *et al.* (2010). Developmental genes identified from the pea aphid draft genome were manually annotated as a part of this study. It was reported that although developmental genes were conserved in *A. pisum*, lineage-specific gene duplication and losses were reported. Specifically, duplications were observed for gene regulators of EGF signaling; receptors and transcription factors belonging to the JAK/STAT signaling pathway; gene families regulating TGFβ signaling; and downstream effectors of Wnt signaling pathways. Duplication of *Ago-1* and *Dcr-1* genes belonging to the small RNA (sRNA) biogenesis pathway was also reported in pea aphids and their differential expression in reproductive morphs was studied (Ortiz-Rivas *et al.*, 2012). Similarly, tissue-specific expression of piRNA biogenesis genes was reported during embryogenesis, implicating their role in phenotypic plasticity (Lu *et al.*, 2011). Transcriptional signatures associated with embryonic development were explored using a microarray-based analysis of gene expression in the late-stage embryos and the early nymphs of the pea aphid (Rabatel *et al.*, 2013). This study revealed the up-regulation of the tyrosine biosynthetic genes along with abundance of cuticular gene expression during pea aphid embryo development. Cuticular proteins have a well-established role in phenotypic plasticity and are now implicated in regulating the development of nymphs to mature adults. It was speculated that up-regulation of a tyrosine metabolic pathway also impacts symbiotic metabolism as this pathway is shared by both the aphid and its endosymbiont.

Another unique feature of aphids is the presence of novel cells called bacteriocytes that house the obligate endosymbiont *Buchnera aphidicola*. Whole genome sequencing of both the pea aphid and its obligate endosymbiont have revealed the close association between the two organisms, with *Buchnera* providing the aphid with amino acids that the aphid cannot synthesize from its sugar-rich diet of phloem sap (Baumann *et al.*, 1995; Hansen and Moran, 2011). Whole-genome analysis of *Buchnera* revealed a markedly reduced genome, with many essential genes required for performing common bacterial functions being absent from the genome. Bacteriocytes isolated from the pea aphid were investigated by EST analysis and qRT-PCR, in order to verify the role of the host in regulating symbiont performance (Nakabachi *et al.*, 2005). Genes encoding amino acid metabolism required for both *Buchnera* and its insect host were highly expressed. Additionally, genes related to mitochondrial and vesicular transport and those encoding for bacterial cell wall degradation were abundant in the bacteriocytes (Nakabachi *et al.*, 2005). Whole genome analysis of the pea aphid revealed several genes with no homology to existing genes from other insect species. It was hypothesized that these genes might be of *Buchenera* origin or play some role in the symbiont association. To validate this hypothesis, RNA-seq was performed with bacteriocytes dissected from the pea aphid (Shigenobu and Stern, 2013). Gene transcripts involved in amino acid metabolism were highly expressed; however, transcripts derived from other non-*Buchnera* bacteria were also abundantly expressed in the bacteriocyte. Interestingly, bacteriocyte-exclusive genes encoded for small cysteine-rich proteins with signal peptides and their expression pattern coincided with the incorporation of the symbionts into the developing embryos and was maintained through all of the life stages (Shigenobu and Stern, 2013). Further studies have revealed that the expression of conserved sRNA can regulate *Buchnera* gene expression (Hansen and Degnan, 2014). Comparative analysis of the sRNA profile of two different life stages of *Buchnera* identified 90 differentially expressed sRNAs, a majority of which were up-regulated in aphid embryos with the predicted role in arginine biosynthesis (Thairu *et al.*, 2018). In addition to *Buchnera*, pea aphids harbor facultative symbionts such as *Serratia* that provide tolerance to heat stress. However, transcriptomic profiling did not reveal significant changes in pea aphid transcriptome in response to *Serratia* (Burke and Moran, 2011).

The successful colonization of aphids depends on their ability to locate and exploit suitable host plants and to adapt to novel environments. In these processes, visual, olfactory, and gustatory cues play important roles for dispersing aphids to locate suitable host plants and initiate phloem sap ingestion. To elucidate the genetics associated with adaptive divergence of pea aphids, a large-scale gene expression study was conducted on clonal populations of pea aphids raised on different host plants (Eyres *et al.*, 2016). Aphids collected from different host plants were cultured on *Vicia faba* under laboratory conditions to raise clonal populations. Clones were transferred to their original host species and new *V. faba* plants. After 2 weeks, these aphids were collected and their heads were dissected and used for preparing RNA-seq libraries. Changes in aphid gene expression in response to host plants were recognized and extreme transcriptomic alterations were observed for aphids transferred to phylogenetically distant host plants. Most of the genes that responded to host diversity were associated with chemosensory and gustatory functions, previously identified in aphid saliva. Several other confirmatory studies using transcriptomic approaches for investigating the roles of chemosensory, odorant binding, and olfactory proteins in host plant selection have also confirmed this phenomenon (Smadja *et al.*, 2009). Initial studies used *in silico* prediction tools to identify 15 genes encoding odorant binding proteins (OBPs) in pea aphid (Xu *et al.*, 2009; Zhou *et al.*, 2014). A follow-up study comparing winged and wingless morphs used qRT-PCR as well as immunolocalization experiments to study the expression pattern of five OBPs (OBP1, OBP3, OBP6, OBP7, OBP8) during different developmental stages of pea aphids (De Biasio *et al.*, 2015).

In response to aphid colonization and phloem feeding, host plants evoke an array of defense responses; however, aphids have evolved to suppress these plant responses by injecting effector proteins through their saliva. A catalogue of transcripts encoding effector proteins was created by both proteomic and transcriptomic profiling of pea aphid salivary glands (Carolan *et al.*, 2011). A total of 1557 transcripts were identified that encoded for 925 proteins, of which approximately one-third were secretory. About half of these proteins were exclusively reported from aphid saliva with no known functions, whereas the remaining proteins showed homology to known effector proteins characterized in plant–nematode interactions (Carolan *et al.*, 2011). Aphid saliva also plays an important role in acquiring and transmitting plant pathogenic viruses during phloem interactions. Virions acquired by aphids move through the gut and salivary tissues before being transmitted through the saliva (Brault *et al.*, 2010). To determine the effect of virus transport on aphid metabolism, a large-scale comparative transcriptomic study was performed in pea aphids inoculated with pea enation mosaic virus (PEMV) (Brault *et al.*, 2010). It was reported that only a minor percentage (1.8%) of the total transcripts regulating vesicular transport were differentially expressed in response to the virus.

4.2.3 Non-model aphids: current advances in transcriptomic research

Understanding genetic differences in closely related species or within the same species can reveal the basis for this taxonomic and phenotypic diversity, respectively. In aphids, reproductive plasticity, winged dimorphism, color polyphenism, and host divergence are frequently observed within the same clonal population and/or in closely related biotypes. Determining the molecular motifs associated with these diverse phenotypes can be achieved by targeted transcriptomic profiling in these species. Although regarded as economically important pests, often genomic resources or information for these aphid species are unavailable or poorly curated. Transcriptome profiling in such 'non-model' aphid species is a necessary step towards building the foundation of advanced genomic research for these species. One of the main advantages of using an NGS platform for exploratory transcriptomic survey is that it requires no prior genomic information, and in recent years this has contributed to rapid development of genomic resources from non-model aphid species (http://1kite.org/). A comprehensive list of transcriptomic surveys and functional annotations conducted in non-model aphid species is provided in Table 4.1.

4.2.3.1 Green peach aphid: model organism for studying insecticide resistance

Myzus persicae (green peach aphid) is a cosmopolitan and polyphagous aphid that is widely studied for its diverse traits in relation to host specificity, color dimorphism, and, most importantly, for insecticide resistance (van Emden and Harrington, 2007). Several lines of evidence for insecticide resistance in aphids have used the green peach aphid resistant genotype as the model system (Puinean *et al.*, 2010). Initial transcriptomic assessment of insecticide

Table 4.1. Transcriptomic studies in non-model aphid species.

Species	Genomic tool	Research objective	Citation
Myzus persicae (green-peach aphid)	ESTs, SNP, microarray	Develop species-specific genomic resources	Ramsey *et al.*, 2007
	Annotated ESTs	Xenobiotic detoxification	Figueroa *et al.*, 2007
	Transcriptome profiling	Insecticide resistance	Silva *et al.*, 2012
	Microarray	Effect of water stress	Vickers, 2012
	Transcriptome analysis	Developmental studies and inter-species divergence	Ji *et al.*, 2016
	De novo transcript assembly from aphid head, aphid body	Identify candidate effector transcripts	Thorpe *et al.*, 2016
	Transcriptome	Transcriptional plasticity related to host diversity	Mathers *et al.*, 2017
	Nanopure sequencing	Detection of plant pathogens from insect vectors	Bronzato *et al.*, 2018
Myzus persicae nicotianae (tobacco aphid)	Microarray, real-time PCR	Insecticide resistance	Cabrera-Brandt *et al.*, 2014
Myzus cerasi (black cherry oat aphid)	*De novo* transcript assembly from aphid head, aphid body	Identify candidate effector transcripts	Thorpe *et al.*, 2016
Aphis craccivora (cow-pea aphid)	Transcriptome sequencing	Mitochondrial gene analysis to understand aphid evolution	Song *et al.*, 2016
Aphis nerii (milkweed-oleander aphid)	Gene expression analysis	Develop species-specific genomic resources	Birnbaum *et al.*, 2018
Aphis glycines (soybean aphid)	454 Transcript sequencing	Develop species-specific genomic resources	Bai *et al.*, 2010
	Transcript profiling of whole aphids and aphid guts	Endosymbiont association	Liu *et al.*, 2012
	Transcriptome profiling	Host-plant resistance interactions	Bansal *et al.*, 2014
	RNA-seq	Response to abiotic and biotic stressors	Enders *et al.*, 2014

Continued

Table 4.1. Continued

Species	Genomic tool	Research objective	Citation
	In silico analysis of RNA-seq data, manual annotation	Identify chemoreceptors and OBP	Robertson *et al.*, 2019
Aphis gossypii (cotton-melon aphid)	Transcriptome analysis of sexual and asexual morphs	Reproductive polyphenism	Liu *et al.*, 2014
	Transcriptome sequencing of insecticide-resistant aphid strains	Thiamethoxam-resistance adaptation	Pan *et al.*, 2015a
	Small RNA transcriptome sequencing	Host-plant resistance	Sattar *et al.*, 2012
	Transcriptome sequencing of insecticide-resistant aphid strains	Genetics of Spirotetramat resistance adaptation, symbiotic bacterial populations associated with Spirotetramat resistance	Pan *et al.*, 2015b Zhang *et al.*, 2016 Xi *et al.*, 2016
Rhopalosiphum padi (bird cherry oat aphid)	*De novo* transcript assembly from aphid head, aphid body	Identify candidate effector transcripts	Thorpe *et al.*, 2016
	Functional annotation	Identification of insecticide-resistance GST-genes	Balakrishnan *et al.*, 2019
Sitobion avenae (grain aphid)	Transcriptome	Evolutionary divergence of two closely related aphid species	Wang *et al.*, 2014
	De novo transcript assembly	Identification of RNAi targets in insect gut	Zhang *et al.*, 2013
	Transcriptome analysis of salivary glands	Identification of potential effectors in aphid saliva	Zhang *et al.*, 2017
	Antennal transcriptome	Identify candidate odorant-binding proteins	Xue *et al.*, 2016
	Transcriptome profiling	Comparative analysis of defense genes in response to host-plant resistance	Huang *et al.*, 2019

Continued

Table 4.1. Continued

Species	Genomic tool	Research objective	Citation
	Real-time PCR, dsRNA feeding assay	Identification and functional characterization of ecdysone receptor and ultraspiracle protein gene orthologs	Yan *et al.*, 2016
Diurophis noxia (Russian wheat aphid)	Gut transcriptome	Compare virulent and avirulent aphids	
Schizaphis graminum (Rondani) (green bug)	Transcriptome and differential gene expression (DEG)	Genetics of imidacloprid adaptation and RNAi	Zhang *et al.*, 2019
Macrosiphum euphorbiae (potato-peach aphid)	RNA-seq	Identification of virus	Teixeira *et al.*, 2018

resistance genes in green peach aphid was developed by Figueroa *et al.* (2007). A cDNA library was constructed from a clonal population of green peach aphid maintained in laboratory conditions and yielded 959 ESTs. Among them several transcripts were annotated as detoxification enzymes for plant xenobiotic products (Figueroa *et al.*, 2007). Recently, a large-scale transcriptomic approach has been successful in revealing the molecular basis of insecticide resistance in *M. persicae*. Comparative microarray profiling of three diverse genotypes of *M. persicae* showing anti-cholinesterase resistance was performed (Silva *et al.*, 2012). In each genotype, resistance was due to either reduced insecticide uptake via the cuticle, altered activity of detoxification enzymes, or point mutations in the insecticide target sites (Li *et al.*, 2007). In the insecticide-susceptible variety, several genes encoding for detoxification enzymes, energy metabolism, cuticular proteins, or extracellular transport proteins were abundantly expressed. In contrast, the variety carrying point mutations at insecticide target sides showed up-regulation of fewer genes, but the up-regulated transcripts encoded for detoxification enzymes and cuticular proteins similarly to the susceptible variety. The third variety showing resistance to multiple insecticide groups had the least number of up-regulated transcripts, and none of these transcripts encoded for an insecticide detoxification enzyme (Silva *et al.*, 2012).

4.2.3.2 Cotton-melon aphid: model organism for *R*-gene mediated genetic plant resistance

Aphis gossypii (cotton and melon aphids) are small aphids with great diversity in geographical range, life cycle, and host specificity (van Emden and Harrington, 2007). These aphids are polyphagous and can infest hundreds of species of crops worldwide. Infestation on cucurbits and cotton cause severe damage by direct feeding, transmission of viral pathogens, and production of sooty molds on cotton lint. *A. gossypii* presents a unique opportunity to investigate the genetics of resistance gene (*R*-gene) mediated aphid resistance in melon plants (Dogimont *et al.*, 2014). In melon, *Vat* (virus aphid transmission) provides dual resistance against *A. gossypii* and aphid-transmitted non-persistent mosaic viruses (Villada *et al.*, 2009; Dogimont *et al.*, 2014). *Vat* is a single dominant gene encoding for an approximately 270 kDa protein with three domains – coil-coil (CC); nucleotide binding site (NBS); and leucine-rich repeats (LRR) – and predicted to localize within the cytoplasm (Dogimont *et al.*, 2014). Aphid resistance to *Vat* is exhibited as anti-xenosis (non-preference), antibiosis (reduced fecundity), and plant tolerance (Sattar *et al.*, 2016).

A comparative analysis of sRNA transcriptome of *A. gossypii* from susceptible (*Vat*$^-$) and resistant (*Vat*$^+$) melon cultivars identified 81 conserved miRNAs, 12 aphid-specific miRNAs, and nine novel miRNAs for *A. gossypii* (Sattar *et al.*, 2012). The size distribution patterns of sRNAs were strikingly different in the two aphid libraries. The *Vat*$^-$ aphid library showed a unimodal size distribution highly represented by 22–23 nt miRNAs. In contrast, the *Vat*$^+$ aphid library showed a bimodal size distribution pattern with both 22–23 nt miRNAs and 25–27 nt sized sequences, and the latter mapped to repeat elements from the *A. pisum* genome and were predicted to be piRNA-like elements. Recent publication of the *A. gossypii* genome has provided the opportunity for improving the existing annotation of these sRNA transcriptomes in the future. In addition to the sRNA transcriptome, a *de novo* whole transcriptome analysis is also in progress (unpublished data). Initial annotation (prior to publication of the *A. gossypii* genome) identified 19,000 transcripts, of which 244 transcripts were differentially expressed in response to *Vat*-mediated resistance (unpublished data).

4.2.3.3 Soybean aphid: newly introduced invasive species

Aphis glycines (soybean aphid) is a newly introduced invasive species in North America. First detected in 2000, this aphid is considered as a major pest of soybean grown in the mid-western plains of North America, and requires regular application of foliar insecticides (Liu *et al.*, 2012). Resistant plant cultivars were identified with genetic resistance towards soybean aphid through the *Rag* (Resistance to *Aphis glycines*) gene, immediately followed by reports of aphid biotypes overcoming this resistance (Hill *et al.*, 2009; Kim *et al.*, 2008). Soon after its introduction in North America, a concerted effort was established to study this aphid. The initial genomic resources for this insect were generated by 454 sequencing of ESTs (Bai *et al.*, 2010). A *de novo* transcriptome assembly from whole aphids and guts followed soon after, which also provided insights into the *Buchnera* transcriptome. A total of 400,000 aphid contigs were identified and from which ≈ 17,000 were annotated. The *Buchnera* transcriptome from soybean aphid had 57% homology to pre-existing *Buchnera* transcripts (Liu *et al.*, 2012).

Secondary endosymbionts such as *Wolbachia* were also identified from this transcriptome. Another interesting transcriptomic study identified differentially expressed genes in aphids in response to host-plant resistance (Bansal *et al.*, 2014). Differential expression of 914 genes was observed and with 352 transcripts significantly up-regulated in aphids feeding on the resistant host plant. Annotation of the up-regulated transcripts revealed that most of them functioned as detoxification genes. Among the down-regulated transcripts, several were annotated as salivary effector proteins (Bansal *et al.*, 2014). In addition to host-plant resistance, age-synchronized soybean aphids were subjected to heat and starvation stress in a transcriptome study conducted by Enders *et al.* (2014). This study revealed that transcriptional responses to host-plant resistance, heat stress, and starvation did not overlap. However, both heat and starvation responses resulted in enhanced expression of glycogen biosynthesis and molecular chaperones as well as suppression of bacterial endosymbiont transcripts and aphid cuticular genes (Enders *et al.*, 2014). Most of the differentially expressed transcripts in response to host-plant resistance were annotated as genes with unknown functions. A recent *in silico* study in soybean aphid identified a large number of chemosensory genes (47 odorant receptors, 61 gustatory receptors, 19 ionotropic receptors, and ten OBPs) (Robertson *et al.*, 2019). Interestingly, the number of these chemosensory genes identified in *A. glycines* is higher than in the model species, *Acyrthosiphon pisum*. However, further confirmation by localization studies in chemosensory specific tissues or RNAi-mediated knockdowns will be required for functional validation.

4.2.3.4 Russian wheat aphid: monocot aphid

Diuraphis noxia (Russian wheat aphid, RWA) is an invasive species that originated in Russia and by 1980 was well established as a pest in the wheat and barley growing regions of East Asia, Africa, and North and South America (van Emden and Harrington, 2007). *R*-gene mediated resistance conferred by *Dn* genes was identified against RWA with reports of several *Dn* genes from cereals, including *Dn1-9*, *Dnx*, and *Dny* (Botha *et al.*, 2005). RWAs affect several members of the cereal family and RWA biotypes have been identified from diverse geographical regions and host species. Biotype 1 (RWA1) and Biotype 2 (RWA2), identified from North America, show distinct behavioral differences on the resistant plants. RWA1 is avirulent to *Dn4* plants, whereas RWA2 can feed and survive on the same *Dn4* plants (Burd *et al.*, 2006). Comparative analysis of gut cDNA libraries from RWA1 and RWA2 revealed significant differences between the two biotypes. Over 50% of RWA1 transcripts showed homology to pea aphid transcripts, where only 17% of the RWA2 matched to pea aphid (Anathakrishnan *et al.*, 2014).

Identifying transcripts expressed in the gut tissues was a priority to develop RNAi tools as a control strategy for RWA biotypes. Zhang *et al.* (2013) created a *de novo* transcript assembly from RWA gut tissue transcripts before and after feeding on wheat phloem sap. This study generated about

30,000 unigenes, out of which 5000 were differentially expressed in response to feeding (Zhang *et al.*, 2013). Among the differentially expressed transcripts, a small subset of 16 genes was chosen for developing RNAi assays. dsRNA targeting of five of these 16 gut-specific genes in artificial diets was successful in knocking down gene expression leading to aphid mortality and stunted growth. However, the most efficacious RNAi target transcript identified from this assay did not match known transcripts from other aphid species.

4.3 Future of Transcriptomics in Aphids

This brief snapshot of past and present transcriptomic surveys in different model and non-model aphid species shows how these tools address diverse topics related to basic biology, host specificity, viral transmission, and polyphenism. Most studies reported have focused on species-specific questions concerning ecological adaptations in the taxa. These studies have revealed novel functions of known genes and identified several new genes that have yet to be annotated. The future of transcriptomics is gradually moving to 'single-cell-omics,' and in the context of aphid biology this would have a profound impact in our understanding of the molecular basis of their feeding behaviors and host specificity. The abundance of genetic information that has become available in the 'omics-era' can also contribute to successful genome editing programs for aphids and related species.

References

Anathakrishnan, R., Sinha, D.K., Murugan, M., Zhu, K.Y., Chen, M.-S. *et al.* (2014) Comparative gut transcriptome analysis reveals differences between virulent and avirulent Russian wheat aphids, *Diuraphis noxia*. *Arthropod-Plant Interactions* 8(2), 79–88. DOI: 10.1007/s11829-014-9293-4.

Bai, X., Zhang, W., Orantes, L., Jun, T.-H., Mittapalli, O. *et al.* (2010) Combining next-generation sequencing strategies for rapid molecular resource development from an invasive aphid species, *Aphis glycines*. *PLoS ONE* 5(6), e11370. DOI: 10.1371/journal.pone.0011370.

Balakrishnan, B., Su, S., Zhang, C. and Chen, M. (2019) Identification and functional characterization of two sigma glutathione *S* -transferase genes from bird cherry-oat aphid (Hemiptera: Aphididae). *Journal of Economic Entomology* 112(1), 416–424. DOI: 10.1093/jee/toy316.

Bansal, R., Mian, R.A.R., Mittapalli, O. and Michel, A.P. (2014) RNA-Seq reveals a xenobiotic stress response in the soybean aphid, *aphis glycines*, when fed aphid-resistant soybean. *BMC Genomics* 15(1), 972. DOI: 10.1186/1471-2164-15-972.

Bao, Y.Y., Qu, L.Y., Zhao, D., Chen, L.B., Jin, H.Y. *et al.* (2013) The genome- and transcriptome-wide analysis of innate immunity in the brown planthopper, *Nilaparvata lugens*. *BMC Genomics* 14(1), 160–168. DOI: 10.1186/1471-2164-14-160.

Baumann, P., Lai, C., Baumann, L., Rouhbakhsh, D., Moran, N.A. *et al.* (1995) Mutualistic associations of aphids and prokaryotes: biology of the genus *buchnera*. *Applied and Environmental Microbiology* 61(1), 1–7.

Bian, L., Li, Z.Q., Ma, L., Cai, X.M., Luo, Z.X. *et al.* (2018) Identification of the genes in tea leafhopper, *Empoasca onukii* (Hemiptera: Cicadellidae), that encode odorant-binding proteins and chemosensory proteins using transcriptome analyses of insect heads. *Applied Entomology and Zoology* 53(1), 93–105. DOI: 10.1007/s13355-017-0533-9.

BIPAA (BioInformatic Platform for Agroecosystem Arthropods) (n.d.) Myzus persicae. Available at: http://bipaa.genouest.org/sp/myzus_persicae (accessed 1 March 2019).

Birnbaum, S.S.L., Rinker, D.C. and Abbot, P. (2018) Maintaining biological cultures and measuring gene expression in *Aphis nerii*: a non-model system for plant-insect interactions. *Journal of Visualized Experiments* 138(138), e58044. DOI: 10.3791/58044.

Blackman, R.L. and Eastop, V.F. (2007) Taxonomic issues. In: van Emden, H.F. and Harrington, R. (eds) *Aphids as Crop Pests*. CAB International, Wallingford, UK, pp. 1–17.

Boerjan, B., Cardoen, D., Verdonck, R., Caers, J. and Schoofs, L. (2012) Insect omics research coming of age. *Canadian Journal of Zoology* 90, 440–455.

Botha, A.-M., Li, Y. and Lapitan, N.L.V. (2005) Cereal host interactions with Russian wheat aphid: a review. *Journal of Plant Interactions* 1(4), 211–222. DOI: 10.1080/17429140601073035.

Brault, V., Tanguy, S., Reinbold, C., Le Trionnaire, G., Arneodo, J. *et al.* (2010) Transcriptomic analysis of intestinal genes following acquisition of pea enation mosaic virus by the pea aphid *Acyrthosiphon pisum*. *Journal of General Virology* 91(Pt 3), 802–808. DOI: 10.1099/vir.0.012856-0.

Brisson, J.A., Davis, G.K. and Stern, D.L. (2007) Common genome-wide patterns of transcript accumulation underlying the wing polyphenism and polymorphism in the pea aphid (*Acyrthosiphon pisum*). *Evolution & Development* 9(4), 338–346. DOI: 10.1111/j.1525-142X.2007.00170.x.

Brisson, J.A. (2010) Aphid wing dimorphisms: linking environmental and genetic control of trait variation. *Philosophical Transactions of the Royal Society B: Biological Sciences* 365(1540), 605–616. DOI: 10.1098/rstb.2009.0255.

Bronzato, A.B., Sherman, D., Stone, A., Gopakumar, A., Wilson, V. *et al.* (2018) Nanopure sequencing as a surveillance tool for plant pathogens in plant and insect tissues. *Plant Disease* 102(8), 1648–1652. DOI: 10.1094/PDIS-04-17-0488-RE.

Buchner, P. (1965) *Endosymbiosis of Animals with Plant Micro-organisms*. Wiley, Chichester, UK.

Burd, J.D., Porter, D.R., Puterka, G.J., Haley, S.D. and Peairs, F.B. (2006) Biotypic variation among North American Russian wheat aphid (Homoptera: Aphididae) populations. *Journal of Economic Entomology* 99(5), 1862–1866. DOI: 10.1093/jee/99.5.1862.

Burke, G.R. and Moran, N.A. (2011) Responses of the pea aphid transcriptome to infection by facultative symbionts. *Insect Molecular Biology* 20(3), 357–365. DOI: 10.1111/j.1365-2583.2011.01070.x.

Cabrera-Brandt, M.A., Silva, A.X., Le Trionnaire, G.L., Tagu, D. and Figueroa, C.C. (2014) Transcriptomic responses of the aphid Myzus persicae nicotianae Blackman (Hemiptera: Aphididae) to insecticides: analyses in the single Chilean clone of the tobacco aphid. *Chilean Journal of Agricultural Research* 74(2), 191–199. DOI: 10.4067/S0718-58392014000200010.

Carolan, J.C., Caragea, D., Reardon, K.T., Mutti, N.S., Dittmer, N. *et al.* (2011) Predicted effector molecules in the salivary secretome of the pea aphid (*Acyrthosiphon pisum*): a dual transcriptomic/proteomic approach. *Journal of Proteome Research* 10(4), 1505–1518. DOI: 10.1021/pr100881q.

Carver, M., Gross, G.F. and Woodward, T.E. (1991) Hemiptera (bugs, leafhoppers, cicadas, aphids, scale insects, etc.). In: CSIR Organization (ed.) *The Insects of Australia, a Textbook for Students and Research Workers*. Melbourne University Press, Victoria, Australia, pp. 429–509.

Chen, E.H., Hou, Q.L., Dou, W., Wei, D.D., Yue, Y. *et al.* (2018) RNA-seq analysis of gene expression changes during pupariation in *Bactrocera dorsalis* (Hendel) (Diptera: Tephritidae). *BMC Genomics* 19(1), 693. DOI: 10.1186/s12864-018-5077-z.

Cloonan, N., Forrest, A.R., Kolle, G., Gardiner, B.B., Faulker, G.J. *et al.* (2008) Stem cell transcriptome profiling via massive-scale mRNA sequencing. *Nature Methods* 5(7), 613–619. DOI: 10.1038/nmeth.1223.

Cortés, T., Tagu, D., Simon, J.C., Moya, A. and Martínez-Torres, D. (2008) Sex versus parthenogenesis: a transcriptomic approach of photoperiod response in the model aphid *Acyrthosiphon pisum* (Hemiptera: Aphididae). *Gene* 408(1–2), 146–156. DOI: 10.1016/j.gene.2007.10.030.

De Biasio, F., Riviello, L., Bruno, D., Grimaldi, A., Congiu, T. *et al.* (2015) Expression pattern analysis of odorant-binding proteins in the pea aphid *Acyrthosiphon pisum. Insect Science* 22(2), 220–234. DOI: 10.1111/1744-7917.12118.

DeLay, B., Mamidala, P., Wijeratne, A., Wijeratne, S., Mittapalli, O. *et al.* (2012) Transcriptome analysis of the salivary glands of potato leaf-hopper. Empoasca Fabae. *Journal of Insect Physiology* 58(12), e1626–1634. DOI: 10.1016/j.jinsphys.2012.10.002.

Dogimont, C., Chovelon, V., Pauquet, J., Boualem, A. and Bendahmane, A. (2014) The *Vat* locus encodes for a CC-NBS-LRR protein that confers resistance to *Aphis gossypii* infestation and *A. gossypii* -mediated virus resistance. *The Plant Journal* 80(6), 993–1004. DOI: 10.1111/tpj.12690.

Dolling, W.R. (1991) *The Hempitera. Natural History Museum Publications*. Oxford University Press, Oxford.

Douglas, A.E. (2006) Phloem-sap feeding by animals: problems and solutions. *Journal of Experimental Botany* 57(4), 747–754. DOI: 10.1093/jxb/erj067.

Enders, L.S., Bickel, R.D., Brisson, J.A., Heng-Moss, T.M., Siegfried, B.D. *et al.* (2014) Abiotic and biotic stressors causing equivalent mortality induce highly variable transcriptional responses in the soybean aphid. *G3: Genes, Genomes. Genetics* 5(2), 261–270.

Eyres, I., Jaquiéry, J., Sugio, A., Duvaux, L., Gharbi, K. *et al.* (2016) Differential gene expression according to race and host plant in the pea aphid. *Molecular Ecology* 25(17), 4197–4215. DOI: 10.1111/mec.13771.

Field, L.M. and Devonshire, A.L. (1998) Evidence that the E4 and FE4 esterase genes responsible for insecticide resistance in the aphid *Myzus persicae* (Sulzer) are part of a gene family. *The Biochemical Journal* 330(1), 169–173. DOI: 10.1042/bj3300169.

Figueroa, C.C., Prunier-Leterme, N., Rispe, C., Sepulveda, F., Fuentes-Contreras, E. *et al.* (2007) Annotated expressed sequence tags and xenobiotic detoxification in the aphid *Myzus persicae* (Sulzer). *Insect Science* 14(1), 29–45. DOI: 10.1111/j.1744-7917.2007.00123.x.

Gallot, A., Rispe, C., Leterme, N., Gauthier, J.-P., Jaubert-Possamai, S. *et al.* (2010) Cuticular proteins and seasonal photoperiodism in aphids. *Insect Biochemistry and Molecular Biology* 40(3), 235–240. DOI: 10.1016/j.ibmb.2009.12.001.

Gauthier, J.-P., Legeai, F., Zasadzinski, A., Rispe, C. and Tagu, D. (2007) AphidBase: a database for aphid genomic resources. *Bioinformatics* 23(6), 783–784. DOI: 10.1093/bioinformatics/btl682.

Goodchild, A.J.P. (1966) Evolution of the alimentary canal in the Hemiptera. *Biological Reviews* 41(1), 97–140. DOI: 10.1111/j.1469-185X.1966.tb01540.x.

Gu, S.H., Wu, K.M., Guo, Y.Y., Field, L.M., Pickett, J.A. *et al.* (2013) Identification and expression profiling of odorant binding proteins and chemosensory proteins between two wingless morphs and a winged morph of the cotton aphid *Aphis gossypii* Glover. *PloS ONE* 8(9), e73524. DOI: 10.1371/journal.pone.0073524.

Hansen, A.K. and Moran, N.A. (2011) Aphid genome expression reveals host-symbiont cooperation in the production of amino acids. *Proceedings of the National Academy of Sciences of the United States of America* 108(7), 2849–2854. DOI: 10.1073/pnas.1013465108.

Hansen, A.K. and Degnan, P.H. (2014) Widespread expression of conserved small RNAs in small symbiont genomes. *The ISME Journal* 8(12), 2490–2502. DOI: 10.1038/ismej.2014.121.

Hasegawa, D.K., Chen, W., Zheng, Y., Kaur, N., Wintermantel, W.M. *et al.* (2018) Comparative transcriptome analysis reveals networks of genes activated in the whitefly *Bemisia tabaci* when fed on tomato plants infected with Tomato yellow leaf curl virus. *Virology* 513(1), 52–64. DOI: 10.1016/j.virol.2017.10.008.

Hill, C.B., Kim, K.-S., Crull, L., Diers, B.W. and Hartman, G.L. (2009) Inheritance of resistance to the soybean aphid in soybean PI 200538. *Crop Science* 49(4), 1193. DOI: 10.2135/cropsci2008.09.0561.

Hornett, E.A. and Wheat, C.W. (2012) Quantitative RNA-Seq analysis in non-model species: assessing transcriptome assemblies as a scaffold and the utility of evolutionary divergent genomic reference species. *BMC Genomics* 13(1), 361. DOI: 10.1186/1471-2164-13-361.

Huang, X., Liu, D., Zhang, R. and Shi, X. (2019) Transcriptional responses in defense-related genes of *Sitobion avenae* (Hemiptera: Aphididae) feeding on wheat and barley. *Journal of Economic Entomology* 112(1), 382–395. DOI: 10.1093/jee/toy329.

Hunter, W.B., Dang, P.M., Bausher, M.G., Chaparro, J.X., McKendree, W. *et al.* (2003) Aphid biology: expressed genes from alate *Toxoptera citricida*, the brown citrus aphid. *Journal of Insect Science* 3(23), 23–7. DOI: 10.1673/031.003.2301.

Ioannidis, P., Lu, Y., Kumar, N., Creasy, T., Daugherty, S. *et al.* (2014) Rapid transcriptome sequencing of an invasive pest, the brown marmorated stink bug *Halyomorpha halys*. *BMC Genomics* 15(1), 1–22. DOI: 10.1186/1471-2164-15-738.

Ji, R., Wang, Y., Cheng, Y., Zhang, M., Zhang, H.-B. *et al.* (2016) Transcriptome analysis of green peach aphid (*Myzus persicae*): insight into developmental regulation and inter-species divergence. *Frontiers in Plant Science* 7, 1562. DOI: 10.3389/fpls.2016.01562.

Ji, R., Yu, H., Fu, Q., Chen, H., Ye, W. *et al.* (2013) Comparative transcriptome analysis of salivary glands of two populations of rice brown planthopper, *Nilaparvata lugens*, that differ in virulence. *PLoS ONE* 8(11), e79612. DOI: 10.1371/journal.pone.0079612.

Karatolos, N., Pauchet, Y., Wilkinson, P., Chauhan, R., Denholm, I. *et al.* (2011) Pyrosequencing the transcriptome of the greenhouse whitefly, *Trialeurodes vaporariorum* reveals multiple transcripts encoding insecticide targets and detoxifying enzymes. *BMC Genomics* 12(56). DOI: 10.1186/1471-2164-12-56.

Kim, K.-S., Hill, C.B., Hartman, G.L., Mian, M.A.R. and Diers, B.W. (2008) Discovery of soybean aphid biotypes. *Crop Science* 48(3), 923–928. DOI: 10.2135/cropsci2007.08.0447.

Kindlmann, P. and Dixon, A. (1989) Developmental constraints in the evolution of reproductive strategies: telescoping of generations in parthenogenetic aphids. *Functional Ecology* 3(5), 531–537. DOI: 10.2307/2389567.

Koch, K.G., Chapman, K., Louis, J., Heng-Moss, T. and Sarath, G. (2016) Plant tolerance: a unique approach to control hemipteran pests. *Frontiers in Plant Science* 7(33), 1363. DOI: 10.3389/fpls.2016.01363.

Le Trionnaire, G., Jaubert-Possamai, S., Bonhomme, J., Gauthier, J.-P., Guernec, G. *et al.* (2012) Transcriptomic profiling of the reproductive mode switch in the pea aphid in response to natural autumnal photoperiod. *Journal of Insect Physiology* 58(12), 1517–1524. DOI: 10.1016/j.jinsphys.2012.07.009.

Le Trionnaire, G., Jaubert, S., Sabater-Muñoz, B., Benedetto, A., Bonhomme, J. *et al.* (2007) Seasonal photoperiodism regulates the expression of cuticular and signalling protein genes in the pea aphid. *Insect Biochemistry and Molecular Biology* 37(10), 1094–1102. DOI: 10.1016/j.ibmb.2007.06.008.

Le Trionnaire, G., Francis, F., Jaubert-Possamai, S., Bonhomme, J., De Pauw, E. *et al.* (2009) Transcriptomic and proteomic analyses of seasonal photoperiodism in the pea aphid. *BMC Genomics* 10(1), 456. DOI: 10.1186/1471-2164-10-456.

Legeai, F., Rizk, G., Walsh, T., Edwards, O., Gordon, K. *et al.* (2010) Bioinformatic prediction, deep sequencing of microRNAs and expression analysis during phenotypic plasticity in the pea aphid, *Acyrthosiphon pisum*. *BMC Genomics* 11(1), 281. DOI: 10.1186/1471-2164-11-281.

Li, X.C., Schuler, M.A. and Berenbaum, M.R. (2007) Molecular mechanisms of metabolic resistance to synthetic and natural xenobiotics. *Annual Review of Entomology* 52(1), 231–253. DOI: 10.1146/annurev.ento.51.110104.151104.

Lister, R., O'Malley, R.C., Tonti-Filippini, J., Gregory, B.D., Berry, C.C. *et al.* (2008) Highly integrated single-base resolution maps of the epigenome in Arabidopsis. *Cell* 13(3), 523–536. DOI: 10.1016/j.cell.2008.03.029.

Liu, S., Chougule, N.P., Vijayendran, D. and Bonning, B.C. (2012) Deep sequencing of the transcriptomes of soybean aphid and associated endosymbionts. *PLoS ONE* 7(9), e45161. DOI: 10.1371/journal.pone.0045161.

Liu, S.-S., Ye, X.-D., Zhao, Q.-Y., Wang, X.-W., Xia, W.-Q. *et al.* (2014) Transcriptomic analyses reveal the adaptive features and biological differences of guts from two invasive whitefly species. *BMC Genomics* 15(1), 370.

Lowe, R., Shirley, N., Bleackley, M., Dolan, S. and Shafee, T. (2017) Transcriptomics technologies. *PLoS Computational Biology* 13(5), 1–23. DOI: 10.1371/journal.pcbi.1005457.

Lu, H.-ling., Tanguy, S., Rispe, C., Gauthier, J.-P., Walsh, T. *et al.* (2011) Expansion of genes encoding piRNA-associated argonaute proteins in the pea aphid: diversification of expression profiles in different plastic morphs. *PLoS ONE* 6(12), e28051. DOI: 10.1371/journal.pone.0028051.

Mathers, T.C., Chen, Y., Kaithakottil, G., Legeai, F., Mugford, S.T. *et al.* (2017) Rapid transcriptional plasticity of duplicated gene clusters enables a clonally reproducing aphid to colonise diverse plant species. *Genome Biology* 18(1), 27. DOI: 10.1186/s13059-016-1145-3.

Miles, P.W. (1972) The saliva of hemipterans. *Advances in Insect Physiology* 9, 183–255.

Misof, B., Liu, S., Meusemann, K., Peters, R.S., Donath, A. *et al.* (2014) Phylogenomics resolves the timing and pattern of insect evolution. *Science* 346(6210), 763–767. DOI: 10.1126/science.1257570.

Mortazavi, A., Williams, B.A., McCue, K., Schaeffer, L. and Wold, B. (2008) Mapping and quantifying mammalian transcriptomes by RNA-Seq. *Nature Methods* 5(7), 621–628. DOI: 10.1038/nmeth.1226.

Mutti, N.S., Louis, J., Pappan, L.K., Pappan, K., Begum, K. *et al.* (2008) A protein from the salivary glands of the pea aphid, *Acyrthosiphon pisum*, is essential in feeding on a host plant. *Proceedings of the National Academy of Sciences of the United States of America* 105(29), 9965–9969. DOI: 10.1073/pnas.0708958105.

Mutti, N.S., Park, Y., Reese, J.C. and Reeck, G.R. (2006) RNAi knockdown of a salivary transcript leading to lethality in the pea aphid, *Acyrthosiphon pisum*. *Journal of Insect Science* 6(38), 1–7. DOI: 10.1673/031.006.3801.

Nagalakshmi, U., Wang, Z., Waern, K., Shou, C., Raha, D. *et al.* (2008) The transcriptional landscape of the yeast genome defined by RNA sequencing. *Science* 20(5881), 1344–1349. DOI: 10.1126/science.1158441.

Nakabachi, A., Shigenobu, S., Sakazume, N., Shiraki, T., Hayashizaki, Y. *et al.* (2005) Transcriptome analysis of the aphid bacteriocyte, the symbiotic host cell that harbors an endocellular mutualistic bacterium, Buchnera. *Proceedings of the National Academy of Sciences of the United States of America* 102(15), 5477–5482. DOI: 10.1073/pnas.0409034102.

Nicholson, S.J., Nickerson, M.L., Dean, M., Song, Y., Hoyt, P.R. *et al.* (2015) The genome of *Diuraphis noxia*, a global aphid pest of small grains. *BMC Genomics* 16(1), 429. DOI: 10.1186/s12864-015-1525-1.

Nussinov, R. (2015) Advancements and challenges in computational biology. *PLoS Computational Biology* 11(1), e1004053. DOI: 10.1371/journal.pcbi.1004053.

Oppenheim, S.J., Baker, R.H., Simon, S. and DeSalle, R. (2014) We can't all be supermodels: the value of comparative transcriptomics to the study of non-model insects. *Insect Molecular Biology* 24(2), 139–154. DOI: 10.1111/imb.12154.

Ortiz-Rivas, B., Jaubert-Possamai, S., Tanguy, S., Gauthier, J.-P., Tagu, D. *et al.* (2012) Evolutionary study of duplications of the miRNA machinery in aphids associated with striking rate acceleration and changes in expression profiles. *BMC Evolutionary Biology* 12(1), 216. DOI: 10.1186/1471-2148-12-216.

Ozsolak, F. and Milos, P.M. (2011) RNA sequencing: advances, challenges and opportunities. *Nature Review Genetics* 12(2), 87–98. DOI: 10.1038/nrg2934.

Pan, Y., Peng, T., Gao, X., Zhang, L., Yang, C. *et al.* (2015a) Transcriptomic comparison of thiamethoxam-resistance adaptation in resistant and susceptible strains of *Aphis gossypii* Glover. *Comparative Biochemistry and Physiology Part D: Genomics and Proteomics* 13, 10–15. DOI: 10.1016/j.cbd.2014.11.001.

Pan, Y., Yang, C., Gao, X., Peng, T., Bi, R. *et al.* (2015b) Spirotetramat resistance adaption analysis of *Aphis gossypii* Glover by transcriptomic survey. *Pesticide Biochemistry and Physiology* 124, 73–80. DOI: 10.1016/j.pestbp.2015.04.007.

Panfilio, K.A. and Angelini, D.R. (2018) By land, air, and sea: hemipteran diversity through the genomic lens. *Current Opinion in Insect Science* 25, 106–115. DOI: 10.1016/j.cois.2017.12.005.

Pitino, M., Coleman, A.D., Maffei, M.E., Ridout, C.J. and Hogenhout, S.A. (2011) Silencing of aphid genes by dsRNA feeding from plants. *PLoS ONE* 6(10), e25709. DOI: 10.1371/journal.pone.0025709.

Puinean, A.M., Foster, S.P., Oliphant, L., Denholm, I., Field, L.M. *et al.* (2010) Amplification of a cytochrome P450 gene is associated with resistance to neonicotinoid insecticides in the aphid *Myzus persicae*. *PLoS Genetics* 6(6), e1000999. DOI: 10.1371/journal.pgen.1000999.

Quan, Q., Hu, X., Pan, B., Zeng, B., Wu, N. *et al.* (2019) Draft genome of the cotton aphid *Aphis gossypii*. *Insect Biochemistry and Molecular Biology* 105, 25–32. DOI: 10.1016/j.ibmb.2018.12.007.

Rabatel, A., Febvay, G., Gaget, K., Duport, G., Baa-Puyoulet, P. *et al.* (2013) Tyrosine pathway regulation is host-mediated in the pea aphid symbiosis during late embryonic and early larval development. *BMC Genomics* 14(1), 235. DOI: 10.1186/1471-2164-14-235.

Ramsey, J.S., Wilson, A.C., de Vos, M., Sun, Q., Tamborindeguy, C. *et al.* (2007) Genomic resources for *Myzus persicae*: EST sequencing, SNP identification, and microarray design. *BMC Genomics* 8(1), 423. DOI: 10.1186/1471-2164-8-423.

Robertson, H.M., Robertson, E.C.N., Walden, K.K.O., Enders, L.S. and Miller, N.J. (2019) The chemoreceptors and odorant binding proteins of the soybean and pea aphids. *Insect Biochemistry and Molecular Biology* 105, 69–78. DOI: 10.1016/j.ibmb.2019.01.005.

Sattar, S., Addo-Quaye, C., Song, Y., Anstead, J.A., Sunkar, R. *et al.* (2012) Expression of Small RNA in *Aphis gossypii* and its potential role in the resistance interaction with melon. *PLoS ONE* 7(11), e48579. DOI: 10.1371/journal.pone.0048579.

Sattar, S., Addo-Quaye, C. and Thompson, G.A. (2016) miRNA-mediated auxin signalling repression during *Vat* -mediated aphid resistance in *Cucumis melo*. *Plant, Cell & Environment* 39(6), 1216–1227. DOI: 10.1111/pce.12645.

Shigenobu, S., Richards, S., Cree, A.G., Morioka, M., Fukatsu, T. *et al.* (2010) A full-length cDNA resource for the pea aphid, *Acyrthosiphon pisum*. *Insect Molecular Biology* 19(Suppl. 2), 23–31. DOI: 10.1111/j.1365-2583.2009.00946.x.

Shigenobu, S. and Stern, D.L. (2013) Aphids evolved novel secreted proteins for symbiosis with bacterial endosymbiont. *Proceedings of the Royal Society B: Biological Sciences* 280(1750), 20121952. DOI: 10.1098/rspb.2012.1952.

Silva, A.X., Jander, G., Samaniego, H., Ramsey, J.S. and Figueroa, C.C. (2012) Insecticide resistance mechanisms in the green peach aphid *Myzus persicae* (Hemiptera: Aphididae) I: A transcriptomic survey. *PLoS ONE* 7(6), 9–11. DOI: 10.1371/journal.pone.0036366.

Smadja, C., Shi, P., Butlin, R.K. and Robertson, H.M. (2009) Large gene family expansions and adaptive evolution for odorant and gustatory receptors in the pea aphid, *Acyrthosiphon pisum*. *Molecular Biology and Evolution* 26(9), 2073–2086. DOI: 10.1093/molbev/msp116.

Song, N., Liang, A.P. and Bu, C.P. (2012) A molecular phylogeny of Hemiptera inferred from mitochondrial genome sequences. *PLoS ONE* 7(11), e48778. DOI: 10.1371/journal.pone.0048778.

Song, N., Zhang, H., Li, H. and Cai, W. (2016) All 37 Mitochondrial genes of aphid *Aphis craccivora* obtained from transcriptome sequencing: implications for the evolution of aphids. *PLoS ONE* 11(6), e0157857. DOI: 10.1371/journal.pone.0157857.

Sseruwagi, P., Wainaina, J., Ndunguru, J., Tumuhimbise, R., Tairo, F. *et al.* (2018) The first transcriptomes from field-collected individual whiteflies (*Bemisia tabaci*, Hemiptera: Aleyrodidae): a case study of the endosymbiont composition. *Gates Open Research* 1, 16. DOI: 10.12688/gatesopenres.12783.3.

Szwedo, J. (2018) The unity, diversity and conformity of bugs (Hemiptera) through time. *Earth and Environmental Science Transactions of the Royal Society of Edinburgh* 107(2-3), 109–128. DOI: 10.1017/S175569101700038X.

Tagu, D., Klingler, J.P., Moya, A. and Simon, J.-C. (2008) Early progress in aphid genomics and consequences for plant–aphid interactions studies. *Molecular Plant-Microbe Interactions* 21(6), 701–708. DOI: 10.1094/MPMI-21-6-0701.

Tagu, D., Prunier-Leterme, N., Legeai, F., Gauthier, J.-P., Duclert, A. *et al.* (2004) Annotated expressed sequence tags for studies of the regulation of reproductive modes in aphids. *Insect Biochemistry and Molecular Biology* 34(8), 809–822. DOI: 10.1016/j.ibmb.2004.05.001.

Teixeira, M.A., Sela, N., Atamian, H.S., Bao, E., Chaudhary, R. *et al.* (2018) Sequence analysis of the potato aphid *Macrosiphum euphorbiae* transcriptome identified two new viruses. *Plos One* 13(3), e0193239. DOI: 10.1371/journal.pone.0193239.

Thairu, M.W., Cheng, S. and Hansen, A.K. (2018) A sRNA in a reduced mutualistic symbiont genome regulates its own gene expression. *Molecular Ecology* 27(8), 1766–1776. DOI: 10.1111/mec.14424.

The International Aphid Genomics Consortium (2010) Genome sequence of the pea aphid *Acyrthosiphon pisum*. *PLoS Biology* 8(2), e1000313. DOI: 10.1371/journal.pbio.1000313.

Thorpe, P., Cock, P.J.A. and Bos, J. (2016) Comparative transcriptomics and proteomics of three different aphid species identifies core and diverse effector sets. *BMC Genomics* 17(1), 172. DOI: 10.1186/s12864-016-2496-6.

Tian, C., Tek Tay, W., Feng, H., Wang, Y., Hu, Y. *et al.* (2015) Characterization of *Adelphocoris suturalis* (Hemiptera: Miridae) transcriptome from different developmental stages. *Scientific Reports* 5(1), 11042. DOI: 10.1038/srep11042.

Tjallingii, W.F. (1985) Electrical nature of recorded signals during stylet penetration by aphids. *Entomologia Experimentalis et Applicata* 38(2), 177–186. DOI: 10.1111/j.1570-7458.1985.tb03516.x.

Tjallingii, W.F. and Hogen Esch, T. (1993) Fine structure of aphid stylet routes in plant tissues in correlation with EPG signals. *Physiological Entomology* 18(3), 317–328. DOI: 10.1111/j.1365-3032.1993.tb00604.x.

Tsuchida, T., Koga, R., Horikawa, M., Tsunoda, T., Maoka, T. *et al.* (2010) Symbiotic bacterium modifies aphid body color. *Science* 330(6007), 1102–1104. DOI: 10.1126/science.1195463.

Tsuchida, T. (2016) Molecular basis and ecological relevance of aphid body colors. *Current Opinion in Insect Science* 17, 74–80. DOI: 10.1016/j.cois.2016.07.005.

van Emden, H.F. and Harrington, R. (eds) (2007) *Aphids as Crop Pests*, 1st edn. CAB International, Wallingford, UK.

Vellichirammal, N.N., Madayiputhiya, N. and Brisson, J.A. (2016) The genomewide transcriptional response underlying the pea aphid wing polyphenism. *Molecular Ecology* 25(17), 4146–4160. DOI: 10.1111/mec.13749.

Vickers, L. (2012) *Aphid Responses to Drought: A Combined Physiological and Transcriptomic Approach*, PhD thesis. University of Birmingham, UK.

Villada, E.S., Gonzalez, E.G., Lopez-Sese, A.I., Castiel, A.F. and Gomez-Guillamon, M.L. (2009) Hypersensitive response to *Aphis gossypii* Glover in melon genotypes carrying the *Vat* gene. *Journal of Experimental Botany* 60(11), 3269–3277. DOI: 10.1093/jxb/erp163.

Wang, Z., Gerstein, M. and Snyder, M. (2009) RNA-Seq: a revolutionary tool for transcriptomics. *Nature Review Genetics* 10(1), 57–63. DOI: 10.1038/nrg2484.

Wang, R., Li, F., Zhang, W., Zhang, X., Qu, C. *et al.* (2017) Identification and expression profile analysis of odorant binding protein and chemosensory protein genes in *Bemisia tabaci* MED by head transcriptome. *PloS ONE* 12(2), e0171739. DOI: 10.1371/journal.pone.0171739.

Wang, D., Liu, Q., Jones, H.D., Bruce, T. and Xia, L. (2014) Comparative transcriptomic analyses revealed divergences of two agriculturally important aphid species. *BMC Genomics* 15(1), 1023. DOI: 10.1186/1471-2164-15-1023.

Wang, X.W., Luan, J.B., Li, J.M., Bao, Y.Y., Zhang, C.X. *et al.* (2010) De novo characterization of a whitefly transcriptome and analysis of its gene expression during development. *BMC Genomics* 11(1), 400. DOI: 10.1186/1471-2164-11-400.

Wenger, J.A., Cassone, B.J., Legeai, F., Johnston, J.S., Bansal, R. *et al.* (2017) Whole genome sequence of the soybean aphid, *Aphis glycines*. *Insect Biochemistry and Molecular Biology*, 102917. DOI: 10.1016/j.ibmb.2017.01.005.

Xi, J., Shang, Q., Gao, X., Pan, Y., Zheng, C. *et al.* (2016) Rapid evolution of symbiotic bacteria populations in spirotetramat-resistant *Aphis gossypii* Glover revealed by pyrosequencing. *Comparative Biochemistry and Physiology Part D: Genomics and Proteomics* 20, 151–158.

Xu, Y.L., He, P., Zhang, L., Fang, S.Q., Dong, S.L. *et al.* (2009) Large-scale identification of odorant-binding proteins and chemosensory proteins from expressed sequence tags in insects. *BMC Genomics* 10(1), 632. DOI: 10.1186/1471-2164-10-632.

Xue, W., Fan, J., Zhang, Y., Xu, Q., Han, Z. *et al.* (2016) Identification and expression analysis of candidate odorant-binding protein and chemosensory protein genes by antennal transcriptome of *sitobion avenae*. *PLoS ONE* 11(8), 1–20. DOI: 10.1371/journal.pone.0161839.

Yan, T., Chen, H., Sun, Y., Yu, X. and Xia, L. (2016) RNA interference of the ecdysone receptor genes EcR and USP in grain aphid (*Sitobion avenae* F.) affects its survival and fecundity upon feeding on wheat plants. *International Journal of Molecular Sciences* 17(12), 2098. DOI: 10.3390/ijms17122098.

Ye, X.-D., Su, Y.-L., Zhao, Q.-Y., Xia, W.-Q., Liu, S.-S. *et al.* (2014) Transcriptomic analyses reveal the adaptive features and biological differences of guts from two invasive whitefly species. *BMC Genomics* 15(1), 370. DOI: 10.1186/1471-2164-15-370.

Zhang, B.-Z., Ma, K.-S., Liu, J.-J., Lu, L.-Y., Chen, X.-L. *et al.* (2019) Differential expression of genes in greenbug (*Schizaphis graminum* Rondani) treated by imidacloprid and RNA interference. *Pest Management Science* 75(6), 1726–1733. DOI: 10.1002/ps.5293.

Zhang, Y., Fan, J., Sun, J., Francis, F. and Chen, J. (2017) Transcriptome analysis of the salivary glands of the grain aphid, *Sitobion avenae*. *Scientific Reports* 7(1), 15911. DOI: 10.1038/s41598-017-16092-z.

Zhang, J., Pan, Y., Zheng, C., Gao, X., Wei, X. *et al.* (2016) Rapid evoluation of symbiotic bacteria populations in spirotetramat-resistant *Aphis gossypii* Glover revealed by pyrosequencing. *Comparative biochemistry and Physiology, Part D20*, 151–158.

Zhang, L., Wang, M.-Y., Li, X.-P., Wang, X.-T., Jia, C.-L. *et al.* (2018) A small set of differentially expressed genes was associated with two color morphs in natural populations of the pea aphid Acyrthosiphon pisum. *Gene* 651, 23–32. DOI: 10.1016/j.gene.2018.01.079.

Zhang, M., Zhou, Y., Wang, H., Jones, H., Gao, Q. *et al.* (2013) Identifying potential RNAi targets in grain aphid (*Sitobion avenae* F.) based on transcriptome profiling of its alimentary canal after feeding on wheat plants. *BMC Genomics* 14(1), 560. DOI: 10.1186/1471-2164-14-560.

Zhang, Y., Fan, J., Sun, J. and Chen, J. (2015) Cloning and RNA interference analysis of the salivary protein C002 gene in *Schizaphis graminum*. *Journal of Integrative Agriculture* 14(4), 698–705. DOI: 10.1016/S2095-3119(14)60822-4.

Zhao, W., Lu, L., Yang, P., Cui, N., Kang, L. *et al.* (2016) Organ-specific transcriptome response of the small brown planthopper toward rice stripe virus. *Insect Biochemistry and Molecular Biology* 70, 60–72. DOI: 10.1016/j.ibmb.2015.11.009.

Zhou, S.S., Sun, Z., Ma, W., Chen, W. and Wang, M.Q. (2014) De novo analysis of the *Nilaparvata lugens* (Stål) antenna transcriptome and expression patterns of olfactory genes. *Comparative Biochemistry and Physiology - Part D: Genomics and Proteomics* 9(1), 31–39.

5 Transcriptomic Research on Honey Bee-Associated Pathogens

JAMES P. TAUBER*

Bee Research Laboratory, Beltsville Agricultural Research Center, US Department of Agriculture, Beltsville, Maryland, USA

5.1 Introduction

The western honey bee (*Apis mellifera; A. mellifera*) is an essential cog in modern agriculture. Honey bees are considered commercial products that are seasonally circulated throughout the USA to pollinate crops. To a lesser extent, bumble bees offer alternative or additive commercial services. Estimates on the annual value of pollination services vary greatly, but on a national level in the USA, beekeepers realized $338 million in income from pollination services (Bond *et al.*, 2014; Ferrier *et al.*, 2018). All things considered, this business is estimated to be approaching $15 billion in value in the USA alone (Calderone, 2012), and hundreds of billions on a global scale (Gallai *et al.*, 2009). The demand for pollinating services appears to be ever-growing as well. This is a purely commercial viewpoint, and bees, notably wild bees (*i.e.*, bees not considered for commercial purposes), are of immeasurable value to the stability and continuation of ecosystems.

Unfortunately, honey bees have many stressors and the resilience of honey bees (as well as that of the beekeepers) is constantly under threat. Stressors include, but are not limited to, overwintering colony losses, pests, pathogens and pesticides (US EPA, 2013). A specific incidence of major colony loss was Colony Collapse Disorder (CCD), which captivated many researchers, beekeepers and the public. CCD is characterized by unexplained symptoms including a sudden loss of adult worker bees from a colony, and was observed in the winter of 2006–2007 when USA beekeepers reported a higher than average loss of their honey bee colonies (Stokstad, 2007; USDA ARS, n.d.). CCD was also observed in other countries (Dainat *et al.*, 2012b). Possible causes of CCD have been attributed to: destroyed foraging land

*james.tauber@usda.gov, jamesptauber@gmail.com

(Naug, 2009; DeGrandi-Hoffman *et al.*, 2010); pesticide/herbicide/antibiotic use (Doublet *et al.*, 2015; Li *et al.*, 2017; Motta *et al.*, 2018); consistent, novel or recombinant pathogens, diseases or pests (vanEngelsdorp *et al.*, 2009); and commercial migration (Welch *et al.*, 2009; Runckel *et al.*, 2011). As of today, the CCD phenomenon has generally subsided, but CCD itself remains somewhat a mystery and annual overwintering colony losses are still observed. Overwintering is a 'distinct physiological and behavioral state' where honey bees have an increased lifespan and remain in the colony during the colder months where they consume stored food (Döke *et al.*, 2015). Although unique and remarkable in and of itself, overwintering is considered a major concern for beekeepers as surveys record an average winter colony loss rate of around 30%. Overwintering success may be affected by pathogens, such as viruses, as well as by pests like *Varroa destructor* (*Varroa* mite) (Highfield *et al.*, 2009; Genersch *et al.*, 2010; Döke *et al.*, 2015). In some cases, pathogens are predictive markers of overwintering loss (Dainat *et al.*, 2012a).

The high value and inherent risk of commercial pollinators has led to hypothesis-driven research that focuses on the impacts of diseases (as well as understanding the pathogens, diseases and pests themselves), land management practices, pesticide use, and other stressors on honey bee health in controlled small-scale experiments, in colony field studies and epidemiologically. Quantitative polymerase chain reaction (qPCR) remains the preferred method to study these aspects on honey bee health and monitoring diseases for diagnostic purposes (Evans, 2006; Evans *et al.*, 2013; Traynor *et al.*, 2016). For example, qPCR is one of many scientific tools that allows us to monitor the abundance of honey bee gene transcripts (*i.e.*, RNA molecules) which are the foundation of biological processes such as the immune system that protects the bee against threats. In return, this data presents, to some extent, how the bee responds to environmental factors. This is also called expression or transcription profiling. Beyond monitoring the bee immune system, scientists can also measure, for example, viral titers in bees from different environments. Although qPCR is instrumental in monitoring differentially regulated gene expression during infection or other stressors, the advent of lower-cost next-generation sequencing offers attractive alternatives to qPCR. This is coupled with a large depository of free publicly available bioinformatic tools (*e.g.*, deposited on github.com), and easier access to next-generation sequencing programs with dedicated servers that would require the researcher to have less experience with bioinformatics (*e.g.*, within the Galaxy environment) (Afgan *et al.*, 2018). Two such transcriptomic methods include microarrays and RNA sequencing (herein, RNA-seq). While microarrays capture many RNA transcripts simultaneously, gene targets are predetermined. RNA-seq is used to sequence all transcripts (to some limitation) without the need to predetermine gene targets (Lowe *et al.*, 2017). RNA-seq is a tool that is especially useful during a discovery phase in order to first identify differentially up- or down-regulated genes between treatments; new pathogens and symbionts; and RNA splicing events and non-coding RNA species (Grozinger and Robinson, 2015). Analysis of RNA-seq data for honey bee studies also benefits from an annotated RNA-seq-improved *Apis mellifera* genome (Elsik *et al.*, 2014). Thus, researchers have a model genome to analyze transcriptomic data (*i.e.*, analyses need not be done *de novo*). Although RNA-seq is now a mainstream application for many projects, microarrays or qPCR may nonetheless still be more appropriate to quickly survey known pathogens.

Pathogens and parasites, and the diseases they cause, are collectively referred to as the pathosphere (Schwarz *et al.*, 2015b), and are considered a constant and emerging threat to the health of honey bees. Shortly after CCD was first recognized (Oldroyd, 2007), a metagenomic study was done to identify possible pathogens and symbionts related to CCD (Cox-Foster *et al.*, 2007). In this study, pathospheric organisms correlated with colony collapse. In addition, they preliminarily identified symbiotic honey bee-associated bacteria. This was followed by a deep RNA-seq study supplemented with qPCR (Cornman *et al.*, 2012b). Using deep RNA-seq, scientists discovered novel viruses and shifts in the microbial population that were associated within colonies showing the symptoms of CCD. In general, it was apparent that no single pathogen was the cause for CCD, but rather a mixture of pathogens with increased abundance and strong pathogen covariance was evident in weakened honey bees.

This chapter will present the use of transcriptomic data in hypothesis-driven research where researchers study the effects of infections, pests and pathogens on honey bee health. The hope is that such data will help us to comprehend phenomena such as CCD and colony loss, and improve both honey bee and wild bee populations.

5.2 Transcriptomics to Study Honey Bee-associated Pathogenic Microbes

Four major honey bee-associated pathogens will be discussed in this chapter section: the microsporidian fungus *Nosema ceranae*, the bacterium *Paenibacillus larvae*, the filamentous fungus *Ascosphaera apis*, and the protozoan *Lotmaria passim*. I will then discuss viruses in the following section.

N. ceranae is a microsporidian fungal parasite originally from the Asian honey bee (*Apis cerana*) that migrated to western countries. This parasite causes nosemosis and it continually replicates intracellularly within intestinal epithelial cells (Chen and Huang, 2010). Research on *N. ceranae* shows that this parasite also modulates gene expression of the host, thus likely improving its success (Antúnez *et al.*, 2009; Martín-Hernández *et al.*, 2017). Two *Nosema* spp. have been described to infect honey bees, *N. ceranae* and *N. apis*, and although it is unclear whether one is outpacing the other worldwide, under laboratory conditions it appears that both equally diminish the lifespan of the honey bee (Natsopoulou *et al.*, 2016). Recently, a new species of *Nosema* was identified, and it remains to be determined how this species affects honey bees (Chemurot *et al.*, 2017).

N. ceranae was one of the pathogens identified in the aforementioned CCD study (Cox-Foster *et al.*, 2007). However, in Cornman *et al.* (2012b), there was no correlation between CCD and *N. ceranae*. By contrast, in observations of CCD colonies in Spain, researchers observed stronger correlations of *N. ceranae* to CCD (Paxton, 2010). Consistent reports show that there is a strong, negative effect of *N. ceranae* on colony success (Higes *et al.*, 2008) as well as on individual success (Higes *et al.*, 2007; Li *et al.*, 2018). Additionally, honey bees from CCD-affected colonies showed that *N. ceranae* is strongly associated with honey bee-associated viruses.

As the demand to control honey bee pathogens increases, various treatments to control nosemosis, from chemicals to genetic interference, have been attempted (Rodríguez-García *et al.*, 2018; Burnham, 2019). There is also much interest in understanding the pathogenicity of *N. ceranae*. Researchers found success in reducing *N. ceranae* spore production using genetic manipulation by small interfering microRNA molecules that target the RNA interference (RNAi) machinery, namely *Dicer*, of *N. ceranae* (Huang *et al.*, 2016). Essentially, these small RNA molecules can selectively inhibit important biological processes in an organism, like *Dicer*. A follow-up study used RNA sequencing to examine the interplay of microRNAs between the host and *N. ceranae* for the reason that *N. ceranae* modulates the host and also because the RNAi machinery of *N. ceranae* appeared vital to its proliferation. Not only was the majority of the *N. ceranae* transcriptome affected by its own RNAi system, but also *N. ceranae* microRNAs modulated honey bee host genes, including a potential apoptosis pathway (Evans and Huang, 2018). Apoptosis is a key honey bee immune defense and is likely to play a role in controlling the intracellular reproduction of *N. ceranae*. This was demonstrated in unaffected bees that had a higher apoptotic index (Higes *et al.*, 2013). As a benefit to the bee, disruption of *Dicer* in *N. ceranae* conversely led to increased expression of two host genes: a putative peritrophic membrane-related gene that builds a protective membrane lining in the intestine, and a cuticular-related gene (apidermin); both of these are physiology-related genes that are important defense genes against intestinal pathogens (Huang *et al.*, 2019). However, *N. ceranae* infection without the use of RNAi to disrupt its proliferation caused a down-regulation of cuticle-related genes as determined in a separate transcriptomic study (Badaoui *et al.*, 2017). Interestingly, in another transcriptomic study, newly emerged bees that were exposed to carbendazim (a fungicide) showed down-regulation of various cuticular genes, including three apidermin genes, which indicated delayed development, but this could have broader impacts such as increased disease (*e.g.*, nosemosis) susceptibility (Wang *et al.*, 2018). Therefore, physical barriers in the host and their manipulation by *N. ceranae* appear to be promising avenues for nosemosis research.

In a separate RNA-seq study to observe differentially regulated genes during nosemosis, scientists found a general decrease in host genes that included melanized encapsulation, immune effectors, and other physiological processes like digestion (Aufauvre *et al.*, 2014). The down-regulation of immune genes was confirmed in one additional RNA-seq experiment (Badaoui *et al.*, 2017). Finally, scientists looked at the genetic regulation of the honey bee brain during nosemosis by RNA-seq. It was found that the Toll and Imd pathways, and their effectors, were neither up- nor down-regulated and likely not mirroring other studies because, unlike viruses, *N. ceranae* does not infect near the brain and is not systemic (Doublet *et al.*, 2016). The overall consensus appears to be that *N. ceranae* eventually suppresses a honey bee immunity defense and modulates host gene expression, which in turn supports its own proliferation (Li *et al.*, 2018).

In a microarray study of pathogens, researchers examined migratory commercial honey bees and whether migratory stress may be related to weakened colonies (Runckel *et al.*, 2011). Honey bees underwent typical commercial practices such as transportation across the United States, antimicrobial treatments and introduction to different food sources. *N. ceranae* was present in all examined colonies yet the researchers could not link migratory practices to nosemosis.

Paenibacillus larvae (*P. larvae*) is a severe bacterial pathogen that causes American foulbrood (AFB) (Genersch, 2010; Graham, 2014). The bacterium devastates honey bee larvae, creating a distinguishable visual characteristic that is used for diagnostics. The disease is so severe that in some countries, laws require burning of the hives once AFB is diagnosed. *P. larvae* is a serious threat that causes high mortality (de Graaf *et al.*, 2013; Krongdang *et al.*, 2019). In Cornman *et al.* (2013), researchers used RNA sequencing to compare transcript reads between healthy and AFB-infected larvae. Researchers found that during infection the larvae mounted a strong antibacterial defense by increasing the expression levels of immune effectors. In a later study it was found that breaching of the protective barrier of the gut (the peritrophic matrix) was a necessary virulence factor for AFB (Garcia-Gonzalez *et al.*, 2014). Yet in the transcriptomic study very few peritrophic matrix genes were up-regulated during infection. More striking was the overall up-regulation of members of the Osiris gene family, which are involved in cuticle formation and have conserved expression across diverse insects during development (Smith *et al.*, 2018). Therefore, transcriptomic data showed that honey bee larvae attempt to mount an immune defense and attempt to repair their cuticle during AFB infection, and therefore revealed factors required by the bacterium to be pathogenic.

Another microbe that affects bee brood is the filamentous fungus *Ascosphaera apis* (*As. apis*), which causes the disease chalkbrood (Jensen *et al.*, 2013). The disease is visually distinguishable because it causes brood to mummify. The disease leads to a reduction in newly emerged bees and can lead to the collapse of a colony as well as possible spillover to other bee species (*e.g.*, bumble bees) (Pereira *et al.*, 2019). Whole transcriptome sequencing was used to examine differentially regulated gene expression between healthy and chalkbrood-affected larvae as well as to confirm the presence of transcripts from the annotated *As. apis* genome in order to identify possible virulence factors of the fungus (Cornman *et al.*, 2012a). One set of enzymes, the chitinases, were identified and proposed to be key virulence factors. For infection, the fungal spores are first consumed by the larvae and enter the intestine, and then the fungus must penetrate the peritrophic matrix and eventually breach the cuticle. *As. apis* is likely utilizing its chitinase repertoire for the latter steps of its pathogenicity. Therefore, the transcriptomic data supports the pathogenicity of the fungus. Also observed during infection were up-regulated *As. apis* genes related to biogenesis of the plasma membrane and the fungal cell wall, thus preliminarily identifying potential RNAi targets to combat chalkbrood. From the transcriptomic studies involving *N. ceranae*, *P. larvae* and *As. apis*, commonalities in pathogenicity came to light and researchers may be able to better combat these diseases with a more generalized, rather than species-specific, approach.

Lotmaria passim (*L. passim*) is a trypanosomatid protozoan that is considered a newly recognized pathogen and is found globally (Ravoet *et al.*, 2015; Schwarz *et al.*, 2015a). *L. passim* causes infection of the honey bee and will elicit a strong immune response that is different from nosemosis (Schwarz and Evans, 2013). Its infection appears strongly correlated with *N. ceranae*, possibly seasonally (Stevanovic *et al.*, 2016), and honey bees under stressed conditions will have a higher infection of *L. passim*. A recent RNA-seq study examined both the honey bee and the *L. passim* transcriptome during a *L. passim* infection; it also monitored mortality and bee physiology (Liu *et al.*, 2019). In agreement with other work, researchers observed an up-regulation of bee immunity genes. More striking, however, was the observation that *L. passim* was altering its transcriptome to cope with the anaerobic and low-nutrient environment of the hindgut, providing interesting insight into how intestinal pathogens are accommodating themselves in their host. The hindgut is also where most of the core symbiotic bacteria reside, and during co-residency *L. passim* infection does not change the bacterial community loads. As *N. ceranae* causes malabsorption in the midgut, this phenomenon appears to also occur during *L. passim* infection as the intestinal epithelial cells showed a starvation response and a suppression of RNA synthesis. A suppression of the latter also suggests that both *L. passim* and *N. ceranae* manipulate the host.

Overall, the use of transcriptomics to study honey bee pathogens, both in controlled infection studies and at the colony level, has provided researchers with valuable insight into host–microbe interactions, key virulence factors and commonalities between pathogenicity.

5.3 Transcriptomics to Study Honey Bee-associated Viruses

Viruses, both persistent and novel (newly discovered or known viruses that recombine), are similarly a threat to the health and prosperity of bees (McMahon *et al.*, 2018; Galbraith *et al.*, 2018; McMenamin and Flenniken, 2018). An important aspect of using the model *A. mellifera* to study insect viruses is that honey bee viruses also have the potential to spill over to wild bee populations (Tehel *et al.*, 2016), and in fact many honey bee-associated viruses are also infectious in other insects (Brettell *et al.*, 2019). Positive single-stranded RNA viruses are the most prevalent honey bee-associated viruses known, and include Israeli acute paralysis virus (IAPV),

Sacbrood virus (SBV), Deformed wing virus (DWV) and Lake Sinai virus (LSV) (McMenamin *et al.*, 2018; Brutscher *et al.*, 2016). These viruses have been linked with either increased mortality, weakened colony health, and/or changes in bee immunity regulation, similar to that from the microbial pathogens in the previous section. From the above-mentioned study by Cox-Foster *et al.* (2007), it was found that IAPV, Kashmir bee virus (KBV) and DWV were more abundant in CCD colonies, although none being exclusively linked to CCD. A deep RNA sequencing experiment on CCD colonies afterwards confirmed that KBV and DWV titers were more abundant in CCD-affected colonies (Cornman *et al.*, 2012b). A microarray study confirmed that CCD colonies consistently had higher pathogen loads, both viral and microsporidial (Johnson *et al.*, 2009). An additional facet of the latter study focused beyond pathogens and found that there was neither a general increase nor decrease in bee immunity gene expression in honey bees from CCD colonies (Johnson *et al.*, 2009). A more recent transcriptomic study (Tozkar *et al.*, 2015), which focused again on weakened colonies versus apparently healthy colonies, showed that, in agreement with the previous CCD studies, failing colonies had a higher loads of viruses. Additionally, in this study, despite the high sensitivity of qPCR, IAPV was only detected in trace amounts by RNA sequencing, thus highlighting one advantage of next-generation sequencing. In the aforementioned study focusing on monitoring pathogen loads of migratory honey bees (Runckel *et al.*, 2011), weak colonies (*i.e.*, colonies with few frames and queen problems) did not correlate with viral loads. Overall, it appears that weak colonies have higher viral titers, but colonies can cope with casual viral loads that occur seasonally. Also, from this study, the researchers discovered four novel viruses by RNA deep sequencing that were omitted from the microarray. One newly discovered virus was Lake Sinai Virus 2 (LSV2). LSV2 and LSV1 (Lake Sinai Virus 1) are of growing interest in honey bee health (Daughenbaugh *et al.*, 2015). LSV2 titers peak at months in or around winter (Runckel *et al.*, 2011) and weak colonies appear to have higher loads of LSV2 and LSV1 (Daughenbaugh *et al.*, 2015). A deep RNA sequencing study confirmed that honey bee colonies can have both LSV2 and LSV1; however, only LSV1 appears to have different abundances between CCD symptomatic and healthy colonies (Cornman *et al.*, 2012b). As noted in Cornman *et al.* (2012b), despite clues provided by sequencing efforts to reveal potentially impactful pathogens correlated to CCD, controlled experiments using pathogen manipulation will be necessary to provide more definitive effects that these viruses may have on honey bees.

There have been numerous controlled experiments whereby honey bees were inoculated with viruses, pathogens, symbionts and/or pesticides in order to monitor changes in immunity, physiology and mortality, and thus to evaluate the effect of viruses and other factors on individual honey bee health. Some of these studies also utilized transcriptomics for analyses. An RNA-seq–epigenetic controlled experiment was performed to study the defense response of the honey bee during IAPV infection (Galbraith *et al.*, 2015). It was found that the Toll immune pathway and RNAi genes were up-regulated during the bee's antiviral defense. This RNA-seq experiment supports an earlier and successful experiment where this virus was manipulated in the host by feeding honey bees with double-stranded RNA (*i.e.*, dsRNA used in RNAi) that targets IAPV (Maori *et al.*, 2009), suggesting that RNAi is an important antiviral defense strategy. DWV is another virus that is found in greater abundance in weakened colonies and is also associated with overwintering losses. It has been discovered that there are three master genotypes of DWV (DWV-A, -B and -C), and each one to some extent is associated with colony loss (Kevill *et al.*, 2017; Tehel *et al.*, 2019; Natsopoulou *et al.*, 2017; Kevill *et al.*, 2019; Brettell *et al.*, 2019). RNA-seq and controlled experiments showed that a recombinant strain of DWV-A and -B can be more virulent than DWV-A itself (Moore *et al.*, 2011; McMahon *et al.*, 2016; Ryabov *et al.*, 2017). A microarray study was done to follow the honey bee transcriptomic response to DWV variants, which also used small RNA libraries to confirm the studied DWV variants (Ryabov *et al.*, 2014). Although researchers observed no clear up-regulation of Toll pathway and RNAi genes (as was the case in the above IAPV study), their prepared and sequenced small RNA library (15–40 nucleotides) showed high relative proportions of sense-stranded 22-nucleotide RNAs, which are the RNAs used by RNAi machinery for RNAi control. Similar to the IAPV study, double-stranded DWV-targeting RNA that was fed to honey bees led to reduced DWV loads (Desai *et al.*, 2012), indicating again that the RNAi pathway is an important antiviral defense. Another transcriptomic study confirmed the above studies in that the RNAi machinery genes and the Toll immune pathway (as well as other immune pathways) were up-regulated during a viral infection (Brutscher *et al.*, 2017); around the same time a separate deep sequencing study confirmed RNAi machinery's role in antiviral defenses (Remnant *et al.*, 2017). An interesting finding from the former study is that any dsRNA (*i.e.*, dsRNA that is not specifically targeting a specific virus) induced an antiviral defense. RNA sequencing and RNA interference studies to understand how honey bees mount an antiviral defense and to monitor (novel) viruses spatially and temporally will be key to curbing the transmission of viruses.

One key aspect of honey bee-associated viruses is that many of them are transmitted by *Varroa destructor*, a parasitic mite that feeds off honey bee fat bodies, which in turn severely impairs young bees and decreases host metabolism (identified through sequencing libraries), all while vectoring many (novel) RNA viruses causing further damage to the colony (Evans and Cook, 2018; Ramsey *et al.*, 2019; Alaux *et al.*, 2011; Levin *et al.*, 2019). As RNAi appears to be a promising tool to combat diseases, this mechanism has been applied to try to combat the mite (Campbell *et al.*, 2016). Using an expressed sequence tag library (another transcriptomic method that is used to identify possible transcripts of interest), researchers confirmed the existence of DWV-B and its recombinant variant in mites, as well as possible gene targets for RNAi to fight the mite (namely, a B-type allatostatin gene and a member of the crustacean hyperglycaemic hormone superfamily gene). Use of RNAi (by double-stranded RNA molecules) targeting these genes confirmed their essential role in the mite whereby a high mortality of mites was seen after dsRNA treatment.

Other RNA transmission routes were recently discovered in honey bee food via RNA-seq (Maori *et al.*, 2019a). Major royal jelly protein is the largest constituent of royal jelly, which is a food source that is fed to all younger larvae in the colony. One of the encoded major royal jelly proteins was characterized to bind to RNAs, even non-specifically, and using RNA-seq the researchers found reads of the whole DWV-B genome bound to major royal jelly proteins (Maori *et al.*, 2019b). Conversely, this protein may serve as a route to share RNA species to combat viruses, as it can also bind to double-stranded RNA.

In 2017, a comprehensive overview was published that examined 19 transcriptomic datasets (microarray and RNA-seq) on host–pathogen interactions (Doublet *et al.*, 2017). The goal was to find common patterns in immune defense shared across the studies. They concluded that virus and mite infections led to differentially expressed genes in the Toll pathway, and that apoptosis appeared to be important for *N. ceranae* infections. These agreed with the other studies that were not included in this large dataset study.

5.4 Other Uses of Transcriptomics with Honey Bees

Although not a focus in this chapter, transcriptomic studies have helped researchers study the possible impact of pesticides on honey bee health (Aufauvre *et al.*, 2014; Li *et al.*, 2019; Shi *et al.*, 2017; Snart, 2011; Wang *et al.*, 2018). Cumulatively, it was shown that doses of insecticides and fungicides alter the expression levels of genes involved in immunity, development, metabolism, and microRNAs (non-coding small RNAs that regulate gene expression) in honey bees. This could have far-reaching effects on disease susceptibility. For example, with regard to honey bee-associated pathogenic microbes, a delay in development and physiological defects like reduced cuticle formation could lead to increased susceptibility to *N. ceranae*, *P. larvae* and *As. apis*, as breaching these physical structures are essential for their pathogenicity.

For researchers who are using next-generation sequencing for honey bee research, the US Department of Agriculture (USDA) offers HoloBee, an extensively curated nucleotide sequence database of honey bee-associated pathogens, opportunists and symbionts (https://data.nal.usda.gov/dataset/holobee-database-v20161), hosted at the National Agricultural Library's Ag Data Commons (Evans *et al.*, 2016). The database consists of two parts, HoloBee-Barcode and HoloBee-Mop, with distinct uses. HoloBee-Barcode sequence data is intended for standardized identification of all sequenced viruses, bacteria, fungi, protozoans and metazoans associated with honey bees (currently 86 viruses, 492 bacteria, 41 fungi, four protozoa and 60 metazoa). HoloBee-Mop includes chromosomal, mitochondrial and plasmid genome assemblies in order to provide as much available genomic sequence information possible for members of the honey bee holobiont community (available for two viruses, 55 bacteria, five fungi, one protozoon, and six metazoa). Used in combination with the honey bee genome, HoloBee-Barcode and HoloBee-Mop allow users to remove known microbes and viruses from their sequencing data, allowing unique reads containing potentially novel species or variants to be analyzed. Additionally, honey bee data can be screened against HoloBee-Mop to remove contaminants, thus improving sequence analysis.

5.5 Conclusion

The constant improvement of sequencing technology will undoubtedly be accompanied by improved sensitivity and specificity, as well as reduction of cost and hands-on time. RNA sequencing strategies continue to be helpful in monitoring novel or recombinant strains of viruses, monitoring known pathogens and attempting to find (a)biotic correlations with backyard and commercial bee population losses. Controlled experiments designed to utilize RNA-seq provide insight into how honey bees mount an immune defense and into other physiological processes activated during infection, and will assist researchers in identifying key virulence factors with the eventual goal to develop new tools to combat bee pathogens. Although variations between experiments may lead to different results, aggregated data will be useful to find commonalities. Also, massive sequencing datasets will require the functional characterization of putative genes and their respective proteins to confirm conclusions.

The use of next-generation sequencing will help us to understand the losses and hardships in bee populations. This in turn will advance our awareness, help us to improve the management of lands and chemicals, and to better control diseases. Overall, it appears that the effect of pathogen loads on colony health is still a heightened concern.

Acknowledgements

James Tauber was supported by an appointment to the Agricultural Research Service (ARS) Research Participation Program administered by the Oak Ridge Institute for Science and Education (ORISE) through an interagency agreement between the US Department of Energy (DOE) and the US Department of Agriculture (USDA). ORISE is managed by ORAU (Oak Ridge Associated Universities) under DOE contract number DE-SC0014664. James Tauber was also supported by the ARS Research Associate Program (Class of 2018). USDA is an equal opportunity provider and employer. The author declares that he has no competing interests. All opinions expressed in this paper are the author's and do not necessarily reflect the policies and views of USDA, ARS, DOE, or ORAU/ORISE. Mention of trade names or commercial products in this publication is solely for the purpose of providing specific information and does not imply recommendation or endorsement by the US Department of Agriculture. I thank D. Lopez, J. Evans, J. Tauber, L. Tauber and A. Childers for writing help and the entire BRL crew for support. I also thank D. McMahon and the Federal Institute for Materials Research and Testing (BAM; Berlin, Germany) for support.

References

Afgan, E., Baker, D., Batut, B., van den Beek, M., Bouvier, D. *et al.* (2018) The Galaxy platform for accessible, reproducible and collaborative biomedical analyses: 2018 update. *Nucleic Acids Research* 46(W1), W537–W544. DOI: 10.1093/nar/gky379.

Alaux, C., Dantec, C., Parrinello, H. and Le Conte, Y. (2011) Nutrigenomics in honey bees: digital gene expression analysis of pollen's nutritive effects on healthy and varroa-parasitized bees. *BMC Genomics* 12(1), 496. DOI: 10.1186/1471-2164-12-496.

Antúnez, K., Martín-Hernández, R., Prieto, L., Meana, A., Zunino, P. *et al.* (2009) Immune suppression in the honey bee (*Apis mellifera*) following infection by *Nosema ceranae* (*Microsporidia*). *Environmental Microbiology* 11(9), 2284–2290. DOI: 10.1111/j.1462-2920.2009.01953.x.

Aufauvre, J., Misme-Aucouturier, B., Viguès, B., Texier, C., Delbac, F. *et al.* (2014) Transcriptome analyses of the honeybee response to *Nosema ceranae* and insecticides. *PLoS ONE* 9(3), e91686. DOI: 10.1371/journal.pone.0091686.

Badaoui, B., Fougeroux, A., Petit, F., Anselmo, A., Gorni, C. *et al.* (2017) RNA-sequence analysis of gene expression from honeybees (*Apis mellifera*) infected with *Nosema ceranae*. *PLoS ONE* 12(3), e0173438. DOI: 10.1371/journal.pone.0173438.

Bond, J., Plattner, K. and Hunt, K. (2014) *U.S. Pollination-Services Market: Fruit and Tree Nuts Outlook: Economic Insight*. USDA Economic Research Service, p. FTS-357SA, 6.

Brettell, L.E., Schroeder, D.C. and Martin, S.J. (2019) RNAseq analysis reveals virus diversity within Hawaiian apiary insect communities. *Viruses* 11(5), 397. DOI: 10.3390/v11050397.

Brutscher, L.M., Daughenbaugh, K.F. and Flenniken, M.L. (2017) Virus and dsRNA-triggered transcriptional responses reveal key components of honey bee antiviral defense. *Scientific Reports* 7(1), 6448. DOI: 10.1038/s41598-017-06623-z.

Brutscher, L.M., McMenamin, A.J. and Flenniken, M.L. (2016) The buzz about honey bee viruses. *PLoS Pathogens* 12(8), e1005757. DOI: 10.1371/journal.ppat.1005757.

Burnham, A.J. (2019) Scientific advances in controlling *Nosema ceranae* (Microsporidia) infections in honey bees (*Apis mellifera*). *Frontiers in Veterinary Science* 6, 79. DOI: 10.3389/fvets.2019.00079.

Calderone, N.W. (2012) Insect pollinated crops, insect pollinators and US agriculture: trend analysis of aggregate data for the period 1992–2009. *PLoS ONE* 7(5), e37235. DOI: 10.1371/journal.pone.0037235.

Campbell, E.M., Budge, G.E., Watkins, M. and Bowman, A.S. (2016) Transcriptome analysis of the synganglion from the honey bee mite, *Varroa destructor* and RNAi knockdown of neural peptide targets. *Insect Biochemistry and Molecular Biology* 70, 116–126. DOI: 10.1016/j.ibmb.2015.12.007.

Chemurot, M., De Smet, L., Brunain, M., De Rycke, R. and de Graaf, D.C. (2017) *Nosema neumanni* n. sp. (Microsporidia, Nosematidae), a new microsporidian parasite of honeybees, *Apis mellifera* in Uganda. *European Journal of Protistology* 61, 13–19. DOI: 10.1016/j.ejop.2017.07.002.

Chen, Y.P. and Huang, Z.Y. (2010) *Nosema ceranae*, a newly identified pathogen of *Apis mellifera* in the USA and Asia. *Apidologie* 41(3), 364–374. DOI: 10.1051/apido/2010021.

Cornman, R.S., Bennett, A.K., Murray, K.D., Evans, J.D., Elsik, C.G. *et al.* (2012a) Transcriptome analysis of the honey bee fungal pathogen, *Ascosphaera apis*: implications for host pathogenesis. *BMC Genomics* 13(1), 285. DOI: 10.1186/1471-2164-13-285.

Cornman, R.S., Tarpy, D.R., Chen, Y., Jeffreys, L., Lopez, D. *et al.* (2012b) Pathogen webs in collapsing honey bee colonies. *PLoS ONE* 7(8), e43562. DOI: 10.1371/journal.pone.0043562.

Cornman, R.S., Lopez, D. and Evans, J.D. (2013) Transcriptional response of honey bee larvae infected with the bacterial pathogen *Paenibacillus larvae*. *PLoS ONE* 8(6), e65424. DOI: 10.1371/journal.pone.0065424.

Cox-Foster, D.L., Conlan, S., Holmes, E.C., Palacios, G., Evans, J.D. *et al.* (2007) A metagenomic survey of microbes in honey bee colony collapse disorder. *Science* 318(5848), 283–287. DOI: 10.1126/science.1146498.

Dainat, B., Evans, J.D., Chen, Y.P., Gauthier, L. and Neumann, P. (2012a) Predictive markers of honey bee colony collapse. *PLoS ONE* 7(2), e32151. DOI: 10.1371/journal.pone.0032151.

Dainat, B., vanEngelsdorp, D. and Neumann, P. (2012b) Colony collapse disorder in Europe. *Environmental Microbiology Reports* 4(1), 123–125. DOI: 10.1111/j.1758-2229.2011.00312.x.

Daughenbaugh, K.F., Martin, M., Brutscher, L.M., Cavigli, I., Garcia, E. *et al.* (2015) Honey bee infecting Lake Sinai viruses. *Viruses* 7(6), 3285–3309. DOI: 10.3390/v7062772.

de Graaf, D.C., Alippi, A.M., Antúnez, K., Aronstein, K.A., Budge, G. *et al.* (2013) Standard methods for American foulbrood research. *Journal of Apicultural Research* 52(1), 1–28. DOI: 10.3896/IBRA.1.52.1.11.

DeGrandi-Hoffman, G., Chen, Y., Huang, E. and Huang, M.H. (2010) The effect of diet on protein concentration, hypopharyngeal gland development and virus load in worker honey bees (*Apis mellifera* L.). *Journal of Insect Physiology* 56(9), 1184–1191. DOI: 10.1016/j.jinsphys.2010.03.017.

Desai, S.D., Eu, Y.-J., Whyard, S. and Currie, R.W. (2012) Reduction in deformed wing virus infection in larval and adult honey bees (*Apis mellifera* L.) by double-stranded RNA ingestion. *Insect Molecular Biology* 21(4), 446–455. DOI: 10.1111/j.1365-2583.2012.01150.x.

Doublet, V., Poeschl, Y., Gogol-Döring, A., Alaux, C., Annoscia, D. *et al.* (2017) Unity in defence: honeybee workers exhibit conserved molecular responses to diverse pathogens. *BMC Genomics* 18(1), 207. DOI: 10.1186/s12864-017-3597-6.

Döke, M.A., Frazier, M. and Grozinger, C.M. (2015) Overwintering honey bees: biology and management. *Current Opinion in Insect Science* 10, 185–193. DOI: 10.1016/j.cois.2015.05.014.

Doublet, V., Labarussias, M., de Miranda, J.R., Moritz, R.F.A. and Paxton, R.J. (2015) Bees under stress: sublethal doses of a neonicotinoid pesticide and pathogens interact to elevate honey bee mortality across the life cycle. *Environmental Microbiology* 17(4), 969–983. DOI: 10.1111/1462-2920.12426.

Doublet, V., Paxton, R.J., McDonnell, C.M., Dubois, E., Nidelet, S. *et al.* (2016) Brain transcriptomes of honey bees (*Apis mellifera*) experimentally infected by two pathogens: black queen cell virus and *Nosema ceranae*. *Genomics Data* 10, 79–82. DOI: 10.1016/j.gdata.2016.09.010.

Elsik, C.G., Worley, K.C., Bennett, A.K., Beye, M., Camara, F. *et al.* (2014) Finding the missing honey bee genes: lessons learned from a genome upgrade. *BMC Genomics* 15(1), 86. DOI: 10.1186/1471-2164-15-86.

Evans, J., Schwarz, R. and Childers, A. (2016) HoloBee database v2016.1. Available at: https://data.nal.usda.gov/dataset/holobee-database-v20161 (accessed 19 May 2019).

Evans, J.D. (2006) Beepath: an ordered quantitative-PCR array for exploring honey bee immunity and disease. *Journal of Invertebrate Pathology* 93(2), 135–139. DOI: 10.1016/j.jip.2006.04.004.

Evans, J.D., Schwarz, R.S., Chen, Y.P., Budge, G., Cornman, R.S. *et al.* (2013) Standard methods for molecular research in *Apis mellifera*. *Journal of Apicultural Research* 52(4), 1–54. DOI: 10.3896/IBRA.1.52.4.11.

Evans, J.D. and Huang, Q. (2018) Interactions among host–parasite microRNAs during *Nosema ceranae* proliferation in *Apis mellifera*. *Frontiers in Microbiology* 9, 698. DOI: 10.3389/fmicb.2018.00698.

Evans, J.D. and Cook, S.C. (2018) Genetics and physiology of Varroa mites. *Current Opinion in Insect Science* 26, 130–135. DOI: 10.1016/j.cois.2018.02.005.

Ferrier, P.M., Rucker, R.R., Thurman, W.N. and Burgett, M. (2018) Economic Effects and Responses to Changes in Honey Bee Health. *US Department of Agriculture, Economic Research Service ERR* 246, 54.

Galbraith, D.A., Fuller, Z.L., Ray, A.M., Brockmann, A., Frazier, M. *et al.* (2018) Investigating the viral ecology of global bee communities with high-throughput metagenomics. *Scientific Reports* 8(1), 8879. DOI: 10.1038/s41598-018-27164-z.

Galbraith, D.A., Yang, X., Niño, E.L., Yi, S. and Grozinger, C. (2015) Parallel epigenomic and transcriptomic responses to viral infection in honey bees (*Apis mellifera*). *PLoS Pathogens* 11(3), e1004713. DOI: 10.1371/journal.ppat.1004713.

Gallai, N., Salles, J.-M., Settele, J. and Vaissière, B.E. (2009) Economic valuation of the vulnerability of world agriculture confronted with pollinator decline. *Ecological Economics* 68(3), 810–821. DOI: 10.1016/j.ecolecon.2008.06.014.

Garcia-Gonzalez, E., Poppinga, L., Fünfhaus, A., Hertlein, G., Hedtke, K. *et al.* (2014) *Paenibacillus larvae* chitin-degrading protein PlCBP49 is a key virulence factor in American foulbrood of honey bees. *PLoS Pathogens* 10(7), e1004284. DOI: 10.1371/journal.ppat.1004284.

Genersch, E. (2010) American foulbrood in honeybees and its causative agent, *Paenibacillus larvae*. *Journal of Invertebrate Pathology* 103(Suppl. 1), S10–19. DOI: 10.1016/j.jip.2009.06.015.

Genersch, E., von der Ohe, W., Kaatz, H., Schroeder, A., Otten, C. *et al.* (2010) The German bee monitoring project: a long term study to understand periodically high winter losses of honey bee colonies. *Apidologie* 41(3), 332–352. DOI: 10.1051/apido/2010014.

Graham, S.M. (2014) American foulbrood and its causative agent, *Paenibacillus larvae*, in New Zealand's registered hives and apiaries. MSc thesis. Victoria University of Wellington, New Zealand.

Grozinger, C.M. and Robinson, G.E. (2015) The power and promise of applying genomics to honey bee health. *Current Opinion in Insect Science* 10, 124–132. DOI: 10.1016/j.cois.2015.03.007.

Higes, M., García-Palencia, P., Martín-Hernández, R. and Meana, A. (2007) Experimental infection of *Apis mellifera* honeybees with *Nosema ceranae* (Microsporidia). *Journal of Invertebrate Pathology* 94(3), 211–217. DOI: 10.1016/j.jip.2006.11.001.

Higes, M., Martín-Hernández, R., Botías, C., Bailón, E.G., González-Porto, A.V. *et al.* (2008) How natural infection by *Nosema ceranae* causes honeybee colony collapse. *Environmental Microbiology* 10(10), 2659–2669. DOI: 10.1111/j.1462-2920.2008.01687.x.

Higes, M., Juarranz, Á., Dias-Almeida, J., Lucena, S., Botías, C. *et al.* (2013) Apoptosis in the pathogenesis of *Nosema ceranae* (*Microsporidia*: Nosematidae) in honey bees (*Apis mellifera*). *Environmental Microbiology Reports* 5(4), 530–536. DOI: 10.1111/1758-2229.12059.

Highfield, A.C., El Nagar, A., Mackinder, L.C.M., Noël, L.M.-L.J., Hall, M.J. *et al.* (2009) Deformed wing virus implicated in overwintering honeybee colony losses. *Applied and Environmental Microbiology* 75(22), 7212–7220. DOI: 10.1128/AEM.02227-09.

Huang, Q., Chen, Y., Neumann, P., Li, W. and Evans, J.D. (2016) Effective silencing of dicer decreases spore load of the honey bee parasite *Nosema ceranae*. *Fungal Genomics & Biology* 6(2). DOI: 10.4172/2165-8056.1000144.

Huang, Q., Li, W., Chen, Y., Retschnig-Tanner, G., Yanez, O. *et al.* (2019) Dicer regulates *Nosema ceranae* proliferation in honeybees. *Insect Molecular Biology* 28(1), 74–85. DOI: 10.1111/imb.12534.

Jensen, A.B., Aronstein, K., Flores, J.M., Vojvodic, S., Palacio, M.A. *et al.* (2013) Standard methods for fungal brood disease research. *Journal of Apicultural Research* 52(1), 1–20. DOI: 10.3896/IBRA.1.52.1.13.

Johnson, R.M., Evans, J.D., Robinson, G.E. and Berenbaum, M.R. (2009) Changes in transcript abundance relating to colony collapse disorder in honey bees (*Apis mellifera*). *Proceedings of the National Academy of Sciences of the United States of America* 106(35), 14790–14795. DOI: 10.1073/pnas.0906970106.

Kevill, J.L., de Souza, F.S., Sharples, C., Oliver, R., Schroeder, D.C. *et al.* (2019) DWV-A lethal to honey bees (*Apis mellifera*): a colony level survey of DWV variants (A, B, and C) in England, Wales, and 32 States across the US. *Viruses* 11(5), 426. DOI: 10.3390/v11050426.

Kevill, J.L., Highfield, A., Mordecai, G.J., Martin, S.J. and Schroeder, D.C. (2017) ABC assay: method development and application to quantify the role of three DWV master variants in overwinter colony losses of European honey bees. *Viruses* 9(11), 314. DOI: 10.3390/v9110314.

Krongdang, S., Evans, J.D., Chen, Y., Mookhploy, W. and Chantawannakul, P. (2019) Comparative susceptibility and immune responses of Asian and European honey bees to the American foulbrood pathogen, *Paenibacillus larvae*. *Insect Science* 26(5), 831–842. DOI: 10.1111/1744-7917.12593.

Levin, S., Sela, N., Erez, T., Nestel, D., Pettis, J. *et al.* (2019) New viruses from the ectoparasite mite *Varroa destructor* infesting *Apis mellifera* and *Apis cerana*. *Viruses* 11(2), 94. DOI: 10.3390/v11020094.

Li, W., Chen, Y. and Cook, S.C. (2018) Chronic *Nosema ceranae* infection inflicts comprehensive and persistent immunosuppression and accelerated lipid loss in host *Apis mellifera* honey bees. *International Journal for Parasitology* 48(6), 433–444. DOI: 10.1016/j.ijpara.2017.11.004.

Li, J.H., Evans, J.D., Li, W.F., Zhao, Y.Z., DeGrandi-Hoffman, G. *et al.* (2017) New evidence showing that the destruction of gut bacteria by antibiotic treatment could increase the honey bee's vulnerability to *Nosema* infection. *PLoS ONE* 12(11), 94.

Li, Z., Yu, T., Chen, Y., Heerman, M., He, J. *et al.* (2019) Brain transcriptome of honey bees (*Apis mellifera*) exhibiting impaired olfactory learning induced by a sublethal dose of imidacloprid. *Pesticide Biochemistry and Physiology* 156, 36–43. DOI: 10.1016/j.pestbp.2019.02.001.

Liu, Q., Lei, J., Darby, A.C. and Kadowaki, T. (2019) Honey bee trypanosomatid parasite dynamically changes the transcriptome during the infection of honey bee and modifies the host physiology. *bioRxiv* 529321.

Lowe, R., Shirley, N., Bleackley, M., Dolan, S. and Shafee, T. (2017) Transcriptomics technologies. *PLoS Computational Biology* 13(5), e1005457. DOI: 10.1371/journal.pcbi.1005457.

Maori, E., Garbian, Y., Kunik, V., Mozes-Koch, R., Malka, O. *et al.* (2019a) A transmissible RNA pathway in honey bees. *Cell Reports* 27(7), 1949–1959. DOI: 10.1016/j.celrep.2019.04.073.

Maori, E., Navarro, I.C., Boncristiani, H., Seilly, D.J., Rudolph, K.L.M. *et al.* (2019b) A secreted RNA binding protein forms RNA-stabilizing granules in the honeybee royal jelly. *Molecular Cell* 74(3), 598–608. DOI: 10.1016/j.molcel.2019.03.010.

Maori, E., Paldi, N., Shafir, S., Kalev, H., Tsur, E. *et al.* (2009) IAPV, a bee-affecting virus associated with Colony Collapse Disorder can be silenced by dsRNA ingestion. *Insect Molecular Biology* 18(1), 55–60. DOI: 10.1111/j.1365-2583.2009.00847.x.

Martín-Hernández, R., Higes, M., Sagastume, S., Juarranz, Á., Dias-Almeida, J. *et al.* (2017) Microsporidia infection impacts the host cell's cycle and reduces host cell apoptosis. *PLoS ONE* 12(2), e0170183. DOI: 10.1371/journal.pone.0170183.

McMahon, D.P., Natsopoulou, M.E., Doublet, V., Fürst, M., Weging, S. *et al.* (2016) Elevated virulence of an emerging viral genotype as a driver of honeybee loss. *Proceedings of the Royal Society B: Biological Sciences* 283(1833), 20160811. DOI: 10.1098/rspb.2016.0811.

McMahon, D.P., Wilfert, L., Paxton, R.J. and Brown, M.J.F. (2018) Emerging viruses in bees: from molecules to ecology. *Advances in Virus Research* 101, 251–291.

McMenamin, A.J., Daughenbaugh, K.F., Parekh, F., Pizzorno, M.C. and Flenniken, M.L. (2018) Honey bee and bumble bee antiviral defense. *Viruses* 10(8), 395. DOI: 10.3390/v10080395.

McMenamin, A.J. and Flenniken, M.L. (2018) Recently identified bee viruses and their impact on bee pollinators. *Current Opinion in Insect Science* 26, 120–129. DOI: 10.1016/j.cois.2018.02.009.

Moore, J., Jironkin, A., Chandler, D., Burroughs, N., Evans, D.J. *et al.* (2011) Recombinants between Deformed wing virus and Varroa destructor virus-1 may prevail in *Varroa destructor*-infested honeybee colonies. *The Journal of General Virology* 92(Pt 1), 156–161. DOI: 10.1099/vir.0.025965-0.

Motta, E.V.S., Raymann, K. and Moran, N.A. (2018) Glyphosate perturbs the gut microbiota of honey bees. *Proceedings of the National Academy of Sciences* 115(41), 10305–10310. DOI: 10.1073/pnas.1803880115.

Natsopoulou, M.E., Doublet, V. and Paxton, R.J. (2016) European isolates of the Microsporidia *Nosema apis* and *Nosema ceranae* have similar virulence in laboratory tests on European worker honey bees. *Apidologie* 47(1), 57–65. DOI: 10.1007/s13592-015-0375-9.

Natsopoulou, M.E., McMahon, D.P., Doublet, V., Frey, E., Rosenkranz, P. *et al.* (2017) The virulent, emerging genotype B of Deformed wing virus is closely linked to overwinter honeybee worker loss. *Scientific Reports* 7(1), 5242. DOI: 10.1038/s41598-017-05596-3.

Naug, D. (2009) Nutritional stress due to habitat loss may explain recent honeybee colony collapses. *Biological Conservation* 142(10), 2369–2372. DOI: 10.1016/j.biocon.2009.04.007.

Oldroyd, B.P. (2007) What's killing American honey bees? *PLoS Biology* 5(6), e168. DOI: 10.1371/journal.pbio.0050168.

Paxton, R.J. (2010) Does infection by *Nosema ceranae* cause 'Colony Collapse Disorder' in honey bees (*Apis mellifera*)? *Journal of Apicultural Research* 49(1), 80–84. DOI: 10.3896/IBRA.1.49.1.11.

Pereira, K., Meeus, I. and Smagghe, G. (2019) Honey bee-collected pollen is a potential source of *Ascosphaera apis* infection in managed bumble bees. *Scientific Reports* 9(1), 4241. DOI: 10.1038/s41598-019-40804-2.

Ramsey, S.D., Ochoa, R., Bauchan, G., Gulbronson, C., Mowery, J.D. *et al.* (2019) *Varroa destructor* feeds primarily on honey bee fat body tissue and not hemolymph. *Proceedings of the National Academy of Sciences* 116(5), 1792–1801. DOI: 10.1073/pnas.1818371116.

Ravoet, J., Schwarz, R.S., Descamps, T., Yañez, O., Tozkar, C.O. *et al.* (2015) Differential diagnosis of the honey bee trypanosomatids *Crithidia mellificae* and *Lotmaria passim*. *Journal of Invertebrate Pathology* 130, 21–27. DOI: 10.1016/j.jip.2015.06.007.

Remnant, E.J., Shi, M., Buchmann, G., Blacquière, T., Holmes, E.C. *et al.* (2017) A diverse range of novel RNA viruses in geographically distinct honey bee populations. *Journal of Virology* 91(16). DOI: 10.1128/JVI.00158-17.

Rodríguez-García, C., Evans, J.D., Li, W., Branchiccela, B., Li, J.H. *et al.* (2018) Nosemosis control in European honey bees, *Apis mellifera*, by silencing the gene encoding *Nosema ceranae* polar tube protein 3. *The Journal of Experimental Biology* 221(19), jeb184606. DOI: 10.1242/jeb.184606.

Runckel, C., Flenniken, M.L., Engel, J.C., Ruby, J.G., Ganem, D. *et al.* (2011) Temporal analysis of the honey bee microbiome reveals four novel viruses and seasonal prevalence of known viruses, Nosema, and Crithidia. *PLoS ONE* 6(6), e20656. DOI: 10.1371/journal.pone.0020656.

Ryabov, E.V., Childers, A.K., Chen, Y., Madella, S., Nessa, A. *et al.* (2017) Recent spread of Varroa destructor virus-1, a honey bee pathogen, in the United States. *Scientific Reports* 7(1), 17447. DOI: 10.1038/s41598-017-17802-3.

Ryabov, E.V., Wood, G.R., Fannon, J.M., Moore, J.D., Bull, J.C. *et al.* (2014) A virulent strain of Deformed wing virus (DWV) of honeybees (*Apis mellifera*) prevails after *Varroa destructor*-mediated, or in vitro, transmission. *PLoS Pathogens* 10(6), e1004230. DOI: 10.1371/journal.ppat.1004230.

Schwarz, R.S. and Evans, J.D. (2013) Single and mixed-species trypanosome and microsporidia infections elicit distinct, ephemeral cellular and humoral immune responses in honey bees. *Developmental and Comparative Immunology* 40(3–4), 300–310. DOI: 10.1016/j.dci.2013.03.010.

Schwarz, R.S., Bauchan, G.R., Murphy, C.A., Ravoet, J., de Graaf, D.C. *et al.* (2015a) Characterization of two species of Trypanosomatidae from the honey bee *Apis mellifera*: Crithidia mellificae Langridge and McGhee, and *Lotmaria passim* n. gen., n. sp. *Journal of Eukaryotic Microbiology* 62(5), 567–583. DOI: 10.1111/jeu.12209.

Schwarz, R.S., Huang, Q. and Evans, J.D. (2015b) Hologenome theory and the honey bee pathosphere. *Current Opinion in Insect Science* 10, 1–7. DOI: 10.1016/j.cois.2015.04.006.

Shi, T.-F., Wang, Y.-F., Liu, F., Qi, L. and Yu, L.-S. (2017) Sublethal effects of the neonicotinoid insecticide thiamethoxam on the transcriptome of the honey bees (Hymenoptera: Apidae). *Journal of Economic Entomology* 110(6), 2283–2289. DOI: 10.1093/jee/tox262.

Smith, C.R., Morandin, C., Noureddine, M. and Pant, S. (2018) Conserved roles of Osiris genes in insect development, polymorphism and protection. *Journal of Evolutionary Biology* 31(4), 516–529. DOI: 10.1111/jeb.13238.

Snart, C.J.P. (2011) *Transcriptome analysis of honey bee larvae following neonicotinoid exposure*. 14 December 2011. Available at: http://eprints.nottingham.ac.uk/12333/ (accessed 19 May 2019).

Stevanovic, J., Schwarz, R.S., Vejnovic, B., Evans, J.D., Irwin, R.E. *et al.* (2016) Species-specific diagnostics of *Apis mellifera* trypanosomatids: A nine-year survey (2007–2015) for trypanosomatids and microsporidians in Serbian honey bees. *Journal of Invertebrate Pathology* 139, 6–11. DOI: 10.1016/j.jip.2016.07.001.

Stokstad, E. (2007) The case of the empty hives. *Science* 316(5827), 970–972.

Tehel, A., Brown, M.J. and Paxton, R.J. (2016) Impact of managed honey bee viruses on wild bees. *Current Opinion in Virology* 19, 16–22. DOI: 10.1016/j.coviro.2016.06.006.

Tehel, A., Vu, Q., Bigot, D., Gogol-Döring, A., Koch, P. *et al.* (2019) The two prevalent genotypes of an emerging infectious disease, Deformed wing virus, cause equally low pupal mortality and equally high wing deformities in host honey bees. *Viruses* 11(2), 114. DOI: 10.3390/v11020114.

Tozkar, C.Ö., Kence, M., Kence, M., Huang, Q. and Evans, J.D. (2015) Metatranscriptomic analyses of honey bee colonies. *Frontiers in Genetics* 6, 100. DOI: 10.3389/fgene.2015.00100.

Traynor, K.S., Rennich, K., Forsgren, E., Rose, R., Pettis, J. *et al.* (2016) Multiyear survey targeting disease incidence in US honey bees. *Apidologie* 47(3), 325–347. DOI: 10.1007/s13592-016-0431-0.

US EPA (2013) *Colony Collapse Disorder*. 29 August 2013. US Environment Protection Agency. Available at: https://www.epa.gov/pollinator-protection/colony-collapse-disorder (accessed 14 May 2019).

USDA ARS (n.d.) US Department of Agriculture, Agricultural Research Service. Available at: https://www.ars.usda.gov/oc/br/ccd/index/ (accessed 14 May 2019).

vanEngelsdorp, D., Evans, J.D., Saegerman, C., Mullin, C., Haubruge, E. *et al.* (2009) Colony collapse disorder: a descriptive study. *PLoS ONE* 4(8), e6481. DOI: 10.1371/journal.pone.0006481.

Wang, K., Fan, R.-L., Ji, W.-N., Zhang, W.-W., Chen, X.-M. *et al.* (2018) Transcriptome analysis of newly emerged honeybees exposure to sublethal carbendazim during larval stage. *Frontiers in Genetics* 9. DOI: 10.3389/fgene.2018.00426.

Welch, A., Drummond, F., Tewari, S., Averill, A. and Burand, J.P. (2009) Presence and prevalence of viruses in local and migratory honeybees (*Apis mellifera*) in Massachusetts. *Applied and Environmental Microbiology* 75(24), 7862–7865. DOI: 10.1128/AEM.01319-09.

6 Cytochrome P450s in the Era of Transcriptomics

BERNARDA CALLA,* AND MAY R. BERENBAUM

Department of Entomology, University of Illinois at Urbana-Champaign, USA

6.1 Introduction

Cytochrome P450 (CYP) constitutes a large and hyperdiverse superfamily of enzymatic proteins that are present in organisms across all three kingdoms of life. By binding to iron as a cofactor, CYP enzymes catalyze oxidative reactions, including mono-oxygenation, hydroxylation, epoxidation and dealkylation, among others (Guengerich, 2001). The structural plasticity and diversity of CYP enzymes allow them to be used by living cells in numerous conserved developmental processes, as well as in non-conserved functions that include the production of secondary metabolites and the detoxification of xenobiotics. For those reasons, CYP enzymes are of wide interest and have been studied from the diverse perspectives of molecular biology, pharmacology, genetics, toxicology, agriculture, physiology, and ecology. Research on P450s is abundant and many comprehensive reviews have been published in the last three decades. In this chapter, we focus on the progress of P450 research in insects, particularly with respect to the contributions of next-generation sequencing as a new, accelerating tool for high-throughput discovery. We use as an exemplar past and recent research on species of Lepidoptera, illustrating the advances that have been achieved in the areas of chemical and molecular ecology and evolutionary relationships within herbivores in this insect order.

6.2 Sequencing of P450s Across Years

Although evidence of CYP enzymes had existed since 1955 (Brodie *et al.*, 1955), the first CYP (now CYP2B1) was formally detected in 1964 (Omura

*Corresponding author: calla2@illinois.edu

and Sato, 1964), and its full DNA sequence was not obtained until 1982 (Fujii-Kuriyama *et al.*, 1982). Still pre-dating the invention of the polymerase chain reaction (PCR), the sequencing of the 1473 bp that make up its coding sequence necessitated the use of classical cloning, restriction digestion, piece-by-piece sequencing using gels, and manual calling of bases. Soon after, the sequence of the first cytochrome CYP from an insect was characterized: CYP6A1 from the house fly *Musca domestica* (Feyereisen *et al.*, 1989). Several other researchers followed suit into the molecular biology of insect P450s, with other pioneer P450 sequencing work including CYP4C1, CYP6A2, CYP4C1, CYP6A2, and CYP4D1 from *Drosophila melanogaster* (Bradfield *et al.*, 1991; Gandhi *et al.*, 1992). The first lepidopteran CYP to be sequenced was CYP6B1 from *Papilio polyxenes*, the black swallowtail, in 1992 (Lepidoptera: Papilionidae) (Cohen *et al.*, 1992). The characterization of CYP6B1 and its function highlighted the importance of CYP enzymes in the interaction between insects and their host plants with respect to the detoxification of plant allelochemicals, and the way this interaction has contributed to the evolution and diversification of P450s. The invention of PCR in 1983 undoubtedly accelerated the speed at which CYP enzymes were detected and sequenced, particularly in relatively small animals such as insects, which presented challenges to isolation of sufficient quantities of inducible protein to enable molecular cloning of cDNAs.

Fast-forward 10 years, the first full P450 repertoire (the first CYPome) from an insect was completed (Tijet *et al.*, 2001) as a logical consequence of the completion of the first reference genome assembly for the model species *D. melanogaster* a year earlier (Adams *et al.*, 2000). The initial version of the *D. melanogaster* CYPome had a total of 90 CYPs, with 85 appearing to be functional genes and seven thought to be pseudogenes. Those figures, however, changed in subsequent years, as the genome of *D. melanogaster* was sequenced to deeper levels, and transcriptomic studies added to the sequencing data. Several of the sequences previously deemed pseudogenes turned out to be complete and functional, but at the time of the first genome sequencing version there was probably not enough coverage, or there were too many scaffolding errors. This situation is very common, and it exemplifies how genomics and transcriptomics are rather collectively a dynamic and continuous research process.

Many molecular techniques that were developed three or four decades ago are still widely used for the discovery of P450s, mostly because of their reliability, accessibility, cost, and overall familiarity to researchers; among these methods is the use of expressed sequence tags (ESTs), derived from cDNA libraries that are constructed with cloned fragments, and bacterial artificial chromosomes (BACS), for the sequencing of large inserts. For example, D'alençon *et al.* (2010) studied syntenic conservation in genomic regions in two noctuid lepidopterans (*Helicoverpa armigera* and *Spodoptera frugiperda*) compared with the silkworm *Bombyx mori*, at the time the only lepidopteran with a fully sequenced genome (International Silkworm Genome Consortium, 2008). The study identified an expansion in the CYP9A subfamily in the two noctuid species and delineated the tandem arrangement of CYP9A genes in the genomes of the three species. Those findings were made possible due to the availability of BAC libraries for *H. armigera* and *S. frugiperda*. Sequencing of BAC and EST libraries, or direct PCR amplicon sequencing, still makes use of the Sanger method for sequencing, which, although of high quality, is of very low efficiency, as only stretches of 800–900 bp of high-quality sequence information can be generated on a single reaction.

As increasing numbers of CYP sequences became available, the need for compiling, organizing, and categorizing the growing volume of information became apparent. A system of nomenclature for CYP sequences was proposed in 1987 (Nebert *et al.*, 1987). The proposed system took into account the phylogenetic relationships between the sequences, and it assigned CYP sequences to families and subfamilies based on percentage identity. In 1997, when the sequences of about 750 CYP enzymes were already published, and with whole-genome sequencing projects looming, the nomenclature system had to be revised to adapt to the ever-increasing number and diversity of CYP sequences (Nelson *et al.*, 1993; Nelson, 1998, 2006). A database into which every newly discovered P450 could be deposited and properly named was created in 2009 and named the 'Cytochrome P450 Homepage' (http://drnelson.uthsc.edu/cytochromeP450.html) (Nelson, 2009). After its creation, several other databases that collect data on CYP enzymes were developed. These include the SuperCYP, which aggregates human P450s with drug and drug interactions information (Preissner *et al.*, 2010) (http://bioinformatics.charite.de/supercyp); the Cytochrome P450 Engineering Database (CYPED), which has information on CYP enzymes across all organisms that are also deposited in the NCBI database and includes information about structure and products (https://cyped.biocatnet.de) (Sirim *et al.*, 2009); and a database on CYP allele nomenclature for human sequences (https://www.pharmvar.org). To date, however, insect CYP sequences are still officially named according to the latest rules established by Nelson *et al.* and deposited in the Cytochrome P450 Homepage.

Beyond P450 identification, and within the realm of gene expression quantitation, techniques such as real-time PCR (RT-PCR) and quantitative PCR (qPCR) have been used to study CYP transcripts one or a few transcripts at a time. The first high-throughput gene expression profiling tool and the prototypic method for modern transcriptomics studies was the microarray. Microarray technology made its appearance around the late 1990s to early 2000. These platforms are constructed with DNA, cDNA, or oligonucleotide fragments printed in a surface such as coated glass. They allow for testing several hundreds or thousands of genes on a single mRNA sample by means of hybridization (Fodor *et al.*, 1991; Derisi *et al.*, 1996). Several studies in insects, mostly in model organisms, utilize these platforms to interrogate gene expression changes across diverse tissues and conditions.

For CYP enzymes, whole-genome Affymetrix gene chip microarrays were used to find 22 CYP-coding genes that were significantly down-regulated in female *D. melanogaster* after mating (McGraw *et al.*, 2004). Johnson *et al.* (2009) utilized oligo-based microarrays to inquire into 9867 honey bee (*Apis mellifera*) genes, identifying CYP4G11 as one of the transcripts that increases in abundance in colonies experiencing colony collapse disorder (CCD).

Although commercially produced whole-genome microarrays exist for several model organisms, it is possible to construct custom-made arrays with only the specific genes of interests, a method that became popular to query the CYPomes of several species. For example, David *et al.* (2005), developed a 'detoxification chip' for the malaria-vector mosquito *Anopheles gambiae*, which included a total of 230 gene fragments including glutathione-S-transferases (GSTs), carboxylesterases (COEs), and 103 CYPs. The chip was then used to compare DDT-resistant and DDT-susceptible mosquito strains, and five CYP-coding genes were expressed at higher levels in the resistant strain. For *D. melanogaster*, a custom cDNA microarray was constructed with the 90 *D. melanogaster* CYPs, several GSTs and housekeeping genes. The chip was used with double hybridization to compare field and laboratory-induced resistance to DDT, demonstrating that resistance in field strains is associated with up-regulation of a single P450 (*Cyp6g1*) in response to DDT, whereas laboratory strain resistance is likely conferred by three different CYP enzymes: CYP6G1, CYP6A8, and CYP12D1 (Le Goff *et al.*, 2003). Although they are high-throughput and fairly efficient, microarrays still have the same main disadvantage as the lower-throughput qPCR and RT-PCR, whereby it is a pre-requisite to have both previous sequence information on the genes of interest and high amounts of RNA in order to obtain definitive information. Another potential shortcoming of microarray technology is the possibility of cross-hybridization, a disadvantage that has greater consequences in the case of gene families with multiple members sharing regions of high sequence similarity.

The first next-generation sequencing (NGS) platform was introduced in 2005 by CuraGen: the 454 Life Sciences (later acquired by Roche). At the time, fewer than 5000 CYP enzymes were sequenced and officially named in the database. The 454 platforms relied on emulsion PCR to obtain about 700 megabases of DNA sequence in about 10 hours. In the context of CYP and transcriptomics, the 454 technology was successfully used across different species, helping to transcend the limits usually encountered by non-model insects. For example, 12 full-length CYP enzymes were discovered in the transcriptome of *Zygaena filipendulae*, many of which became candidates for the biosynthesis of cyanogenic glucosides (Zagrobelny *et al.*, 2009), defense chemicals that are widespread in plants that are sequestered by several insect specialists or, in the case of *Z. filipendulae*, also synthesized *de novo*. A life-stage comprehensive *Anopheles funestus* transcriptome was sequenced with 454 technologies (Gregory *et al.*, 2011), and this work helped in the identification of a homolog of the *An. gambiae* CYP4H18 as the most differentially expressed contig in a pyrethroid-resistant strain. Although 454 sequencing has been on the market and used until very recently, ultimately it could not compete with the efficiency of newer technologies.

The first Illumina (then Solexa/Illumina) platform, the 'Illumina Genome Analyzer' (GA), was introduced later in 2005, and its use led to the coining of the broadly used term RNA-seq, which is technically similar to the sequencing-and-counting method used with 454 reads. To date, due to their high-throughput, read quality, and continually decreasing cost, Illumina 'sequencing-by-synthesis' platforms are still the most widely used for transcriptomics. The first transcriptomics experiment using the Illumina GA was published in 2008 and involved *Arabidopsis thaliana* (Lister *et al.*, 2008); human and yeast studies quickly followed (Mortazavi *et al.*, 2008; Nagalakshmi *et al.*, 2008). At that time of introduction of Illumina platforms, the number of sequenced P450s had already doubled since the first use of NGS, and within the past 10 years alone the number of fully sequenced and officially named CYPs has increased more than eight-fold, for a total of about 55,000 CYPs by the end of 2017 (Fig. 6.1). Evidently, obtaining sequence information for CYP enzymes across whole genomes is now a very straightforward process; however, the massive accumulation of data had led to a different set of experimental challenges, including annotation and functional characterization.

To date, at least 130 insect genomes have been fully sequenced (i5kNAL database (https://i5k.nal.usda.gov/species); InsectBase (http://www.insect-genome.com); Yin *et al.*, 2016), and although many of these genomes have gone through a process of automatic – and in several cases manual – annotation, their CYPomes are still not curated. Annotation using automated tools has the capability of identifying sequences that are members of the P450 superfamily, but those tools are usually unable to classify gene models into families or subfamilies, and annotations are based on model organisms, mostly *D. melanogaster* for insect genes. Some automatic tools can separate pseudo-genes and fragments; however, several arthropod genomes have

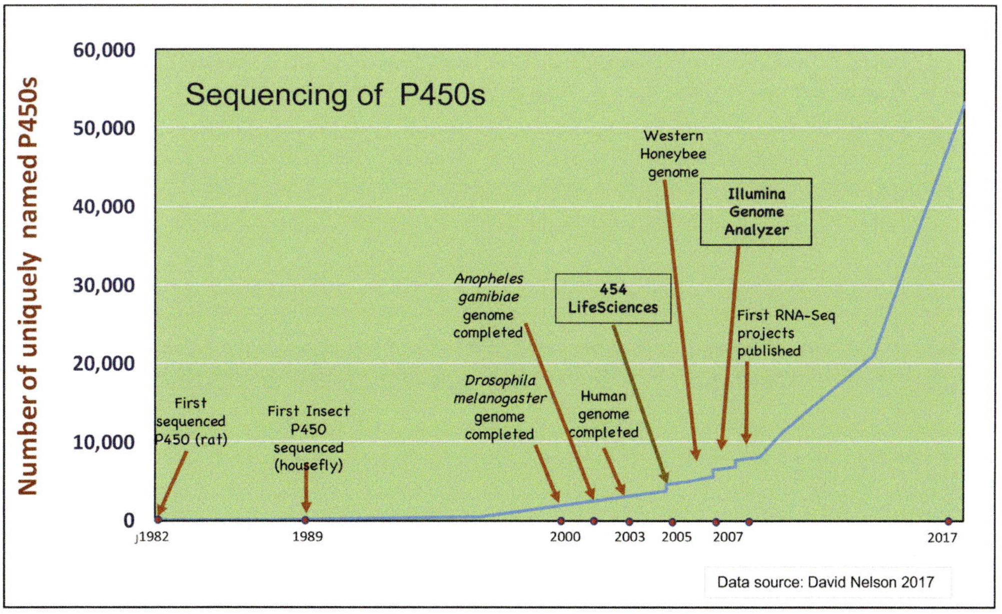

Fig. 6.1. The number of cytochrome P450s sequences has exponentially increased since the introduction of next-generation sequencers. The new challenge is assigning functions to these enzymes. The number of P450s is based in the Cytochrome P450 Homepage database (http://drnelson.uthsc.edu/cytochromeP450.html).

relied on simple Basic Local Alignment Search Tool (BLAST) searches to account for CYP enzymes, greatly over-inflating the actual number of these enzymes in the genomes. For example, 133 P450s coding genes were reported for the genome of the winter moth *Operophtera brumata* (Derks *et al.*, 2015). Upon close inspection, however, only about 102 of these sequences appear to be real P450s (Calla *et al.*, 2017). Another common error found in automatic annotations is the missing of stop codons, making two or more CYP-coding genes appear as a single one when they are arranged into tight clusters, as is the case of many CYPs that originated by gene duplication. An example of this situation is presented for the navel orangeworm genome in Fig. 6.2. The lack of proper CYPome annotation and naming usually makes completed genomes unusable for accurate comparison of CYP repertoires with other arthropods, especially in the context of phylogenomics with species- and gene-tree reconciliations. Examples of fully sequenced and annotated insect CYPomes are scarce compared with the number of fully sequenced insect genomes (Table 6.1).

6.3 Why CYPomes Matter

In contrast to studies focusing on single or several CYP enzymes and their substrates, and with the availability of (or the potential of obtaining) full insect CYPomes, it is now possible to test ecological and evolutionary hypotheses in a more definitive way by utilizing comparative genomics approaches. With the ability to express CYP enzymes heterologously to elucidate their metabolic functions, it became clear that, across plant and insect species, CYP enzymes have been central in shaping terrestrial biodiversity by means of coevolutionary interactions that have fueled reciprocal adaptive radiations in plant allelochemical biosynthesis and insect detoxification genes (reviewed in Omura, 2013). One exemplary case is the P450-dependent production of furanocoumarin compounds in a relatively small number of plant families

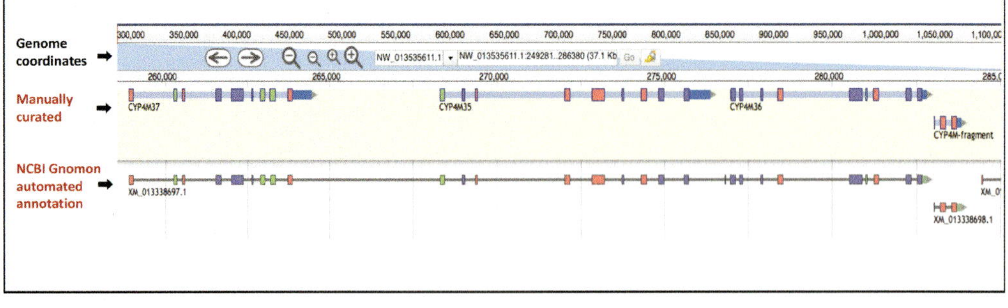

Fig. 6.2. A snapshot of the Web Apollo browser showing the navel orangeworm genome (*Amyelois transitella*, Lepidoptera: Pyralidae) (https://i5k.nal.usda.gov/available-genome-broser). A section of a scaffold is shown where the automated annotation (lower track, NCBI annotation) has merged three P450s (middle track, manually curated). This is an example of the importance of manual curation for accurate downstream analyses of P450s.

Table 6.1. Examples of insect genomes for which the full cytochrome P450 sets (CYPomes) have been annotated and curated. Fully sequenced and annotated insect CYPomes are scarce compared to the number of fully sequenced insect genomes.

Organism	Common name	Authors	Number of P450s
Pediculus humanus	Human body louse	Lee *et al.*, 2010	38
Apis mellifera	Western honey bee	Claudianos *et al.*, 2006	46
Chilo suppressalis	Striped rice stemborer	Wang *et al.*, 2014	77
Depressaria pastinacella	Parsnip webworm	Calla *et al.*, 2017	79
Bombyx mori	Silkworm	Ai *et al.*, 2010	84
Plutella xylostella	Diamondback moth	Yu *et al.*, 2015	84
Nasonia vitripennis	Jewel wasp	Oakeshott *et al.*, 2010	92
Amyelois transitella	Navel orangeworm	Calla *et al.*, 2017	89
Anopheles gambiae	Malaria vector mosquito	Ranson *et al.*, 2002	111
Tribolium castaneum	Red flour beetle	Richards *et al.*, 2008	143

(Berenbaum *et al.*, 1991), and the concomitant P450-dependent detoxification of these compounds in their insect associates (Li *et al.*, 2004; Berenbaum, 2002).

Insect CYP enzymes fall into four different clans: the CYP2, CYP3, CYP4, and mitochondrial clans. The CYP2 and mitochondrial clans are composed of fewer, and generally more conserved, genes that play roles in development and other essential functions such as metabolism of endogenous compounds. These two clans are very stable, with many one-to-one orthologs across species, and rarely show duplications or losses, an indication of their essential functions. The CYP4 and CYP3 clans, in turn, are more variable across species. In Lepidoptera, the CYP3 and CYP4 clans have many more gene members than the CYP2 clan and the mitochondrial clan. The origin of the diversity of CYP3 and CYP4 is hypothesized to be rooted in the fact that enzymes that belong to these groups are implicated in processes of xenobiotic detoxification reviewed in Feyereisen (2011). For these reasons, patterns of diversification that arose in response to interactions with the environment are better seen across families and subfamilies within CYP3 and CYP4 clans.

One such commonly observed diversification pattern in CYPomes is the so-called 'P450 bloom', where within a specific lineage a single CYP subfamily has many more members compared with other subfamilies within the same lineage or species. For insects, these blooms have been found mostly in the CYP3 clan. For example, in the honey bee *A. mellifera,* which has a reduced repertoire of P450s (as well as other detoxificative genes), members of the CYP6AS subfamily outnumber all other *A. mellifera* CYP subfamilies. *A. mellifera* has half the size of the CYPome of the parasitic wasp *Nasonia vitripennis* (46 versus 92 CYP genes), but it has nearly twice as many CYP6AS subfamily members (17 in *A. mellifera* versus nine in *N. vitripennis*) (Claudianos *et al.*, 2006; Oakeshott *et al.*, 2010). Heterologous expression of four of the 17 *A. mellifera* CYP6AS enzymes confirmed that they are capable of metabolizing quercetin, which is widely present in floral nectar and pollen. In this example, CYPome subfamily size is reflective of diversification in the context of specialization (Mao *et al.*, 2009; Johnson *et al.*, 2018).

Within Lepidoptera, at least four subfamilies (i.e. CYP6AB, CYP6AE and CYP6B, and CYP9A) and one family (CYP321) have consistently bloomed in several species. These blooms, rather than random events of duplication and diversification, seem to be highly correlated with diet breadth (Calla *et al.*, 2017). Many studies carried over the past three decades implicated the same blooming families and subfamilies in the metabolism of xenobiotic chemicals. For example, in the tobacco budworm *Heliothis virescens*, CYP9A1 expression patterns are correlated with dithiocarb insecticide resistance (Rose *et al.*, 1995), and CYP9A genes in *Spodoptera litura* are known to be up-regulated by plant allelochemicals and insecticides (Wang *et al.*, 2015). CYP9A20 and CYP9A21 of the silkworm *B. mori* are expressed in the fat body, and CYP9A22 is expressed in the midgut, indicating their possible role in detoxification (Ai *et al.*, 2010). CYP9A2 is induced in the tobacco hornworm *Manduca sexta* by ingestion of artificial diets containing the drug phenobarbital, the wild tomato constituents 2-undecanone and 2-tridecanone, and the non-host linear furanocoumarin xanthotoxin in a concentration- and time-dependent manner, whereas CYP9A4 and CYP9A5 are induced upon consumption of xanthotoxin (Stevens *et al.*, 2000). CYP9A38 of the rice leafroller *Cnaphalocrocis medinalis* was highly expressed in the midgut relative to other tissues (Chen *et al.*, 2015), and several CYP9A from *S. frugiperda* were induced by plant allelochemicals and by pesticides (Giraudo *et al.*, 2015).

The lepidopteran CYP6AE subfamily also illustrates how tandem duplication and diversification within the same subfamily are correlated with functionality. Within the genome of *Depressaria pastincella*, the parsnip webworm, the CYP6AE subfamily is the most diverse, containing 11 different genes with sequence identitities ranging from 53% to 95%. This family is also the second largest in the navel orangeworm *Amyelois transitella* (seven genes). Only two CYP6AE sequences are present in the CYPomes of *Plutella xylostella* and *Chilo suppressalis*. CYP6AE enzymes in *C. medinalis* are involved in detoxification of rice allelochemicals (Liu *et al.*, 2010). CYP6AE14 of *H. armigera* is indirectly involved in gossypol detoxification (Mao *et al.*, 2007; Krempl *et al.*, 2016). *Spodoptera littoralis* CYP6AE47, CYP6AE48, CYP6A49 and CYP6AE50 are expressed at varying levels in larval guts (Pottier *et al.*, 2012). A recent study used CRISPR/Cas9 technology to globally knock out the CYP6AE subfamily in *H. armigera*, resulting in a strain that had significantly higher susceptibility (measured as LC_{50}) to phytochemicals, including xanthotoxin, and insecticides, including esfenvalerate and indoxacarb (Wang *et al.*, 2018).

Many more examples exist of gene-by-gene detection of P450s and their functions, and next-generation sequencing has made it possible to corroborate those previous findings and move studies much further into whole transcriptome comparisons. Techniques such as CRISPR/Cas9 and RNAi depend on the use of whole genome and transcriptome sequencing to identify targets and prevent off-target effects. In the case of CRISPR/Cas9, it is also necessary to have sequence information to identify protospacer adjacent motif (PAM) regions, which are necessary for Cas9 to bind.

6.4 The Future of P450 Sequencing

The past decade has witnessed the explosion of data and expansion of knowledge brought by next-generation sequencing, and, although obtaining sequencing information from genomes and transcriptomes of insects is now accessible to most laboratories, there are still many challenges to overcome, especially for non-model species. One of those challenges is untangling the multitude of confounding factors that short-read, high-throughput sequencing carries. These confounding factors are, for example, heterozygosity, gene copy number variation, alternative spliced isoforms, and allelic variation. These factors are particularly problematical within the context of large gene families with many related members, such as the P450 superfamily. The high sequence similarity at specific regions, coupled with the tandem clustering of several CYPs, makes transcript assembly (for *de novo* transcriptomics) and read mapping (for reference-based transcriptomics) very challenging, as the short reads might not map uniquely, making some transcripts appear more abundant while overlooking others. Although several algorithms have been developed to overcome those challenges, it is still necessary (as stated previously) to subject the CYPomes to exhaustive manual curation to prevent these errors.

Current technologies, including the most widely used Illumina sequencing, involve the fragmenting of genomic DNA or cDNA into hundreds of thousands of fragments that are then ligated to adapters. After sequencing, single or paired reads of 50 and up to 200 nt long are generated, with the size depending on the sequencing method. Most transcriptome studies use 100 or 150 nt long reads, with single-end reads for reference-based transcriptomics or paired-end reads for the *de novo* transcriptome assembly and quantification. These reads are then assembled to generate the best possible representation of the original transcriptome or mapped to a reference genome and counted (for a review, see Korf, 2013). The accuracy with which the task of reconstructing a transcriptome or mapping reads to a reference genome for direct quantitation can be accomplished depends on many factors. For non-model organisms in particular, the main caveat is in the usually lower-quality genome assemblies available and in the genetic variability that exists in field or wild populations that are often the focus of ecological and evolutionary studies. For these reasons, a better approach to assembling and/or complementing full insect transcriptomes, especially when the main aim is to identify and discriminate between different CYP-coding genes and their allelic variants, is available in the newly developed third-generation sequencing (TGS) methods that generate long reads.

Long-read sequencing was achieved only recently. For example, Pacific Biosciences (PacBio) Iso-Seq technology utilizes single-molecule real-time sequencing (SMRT) technology (Levene *et al.*, 2003; Eid *et al.*, 2009), and it can generate reads of up to 10 kb or longer, enough to cover the full length of the majority of transcripts. We have utilized Iso-Seq to refine our understanding of P450 content in the parsnip webworm and found sets of P450s that form chimeric transcripts, where enzymes in the CYP6AB and CYP6AE subfamilies share N-terminal domains (Calla *et al.*, 2017., in preparation), a phenomenon previously found in human P450s (Annalora *et al.*, 2017). Only one other published insect study has made use of Iso-seq technology to obtain a transcriptome of the mitochondria of *Erthesina fullo*, the yellow spotted stink bug (Gao *et al.*, 2016). PacBio sequencing, however, still suffers from low quality and lower throughput compared with Illumina platforms, which increases the cost. At this time, Iso-Seq could be more effectively used in combination with Illumina to obtain more comprehensive data.

Another long-read technology that could be used for transcriptomics is the Oxford Nanopore, which can sequence a full molecule without the need of fluorescent labeling or PCR-amplification (Clarke *et al.*, 2009). This method is relatively inexpensive. Bayega *et al.* (2018) unraveled the isoform complexity in the transcriptome of developing embryos of the olive fruit fly *Bactrocera oleae*. Remarkably, this study discovered 3553 novel genes compared with the most current publicly available *B. oleae* transcriptome in NCBI. This work dramatically illustrates the technological gaps that remain to be filled in order to fully document the transcriptome landscapes of non-model organisms, including, in our case, their CYPomes.

Sequencing of P450s has come a long way since its origins half a century ago and reflects the rapid pace of the history of genomics. We expect that, in the next 5–10 years, CYPomes of all species of Lepidoptera of agricultural, ecological, scientific, and conservation importance will be available and that the data produced from this effort will answer many hitherto refractory biological questions that have defied multiple generations of effort to answer.

Acknowledgement

This work was supported by the National Science Foundation grant # NSFDEB1457731.

References

Adams, M.D., Celniker, S.E., Holt, R.A., Evans, C.A., Gocayne, J.D. *et al.* (2000) The genome sequence of *Drosophila melanogaster*. *Science* 287(5461), 2185–2195. DOI: 10.1126/science.287.5461.2185.

Ai, J., Yu, Q., Cheng, T., Dai, F., Zhang, X. *et al.* (2010) Characterization of multiple CYP9A genes in the silkworm, *Bombyx mori*. *Molecular Biology Reports* 37(3), 1657–1664. DOI: 10.1007/s11033-009-9580-9.

Annalora, A.J., Marcus, C.B. and Iversen, P.L. (2017) Alternative splicing in the cytochrome P450 superfamily expands protein diversity to augment gene function and redirect human drug metabolism. *Drug Metabolism and Disposition* 45(4), 375–389. DOI: 10.1124/dmd.116.073254.

Berenbaum, M.R. (2002) Postgenomic chemical ecology: from genetic code to ecological interactions. *Journal of Chemical Ecology* 28(5), 873–896. DOI: 10.1023/A:1015260931034.

Bradfield, J.Y., Lee, Y.H. and Keeley, L.L. (1991) Cytochrome P450 family 4 in a cockroach: Molecular cloning and regulation by hypertrehalosemic hormone. *Proceedings of the National Academy of Sciences* 88(10), 4558–4562. DOI: 10.1073/pnas.88.10.4558.

Bayega, A., Oikonomopoulos, S., Zorbas, E., Wang, Y.C., Gregoriou, M.-E. *et al.* (2018) Transcriptome landscape of the developing olive fruit fly embryo delineated by Oxford nanopore long-read RNA-seq. *bioRxiv* 478172.

Berenbaum, M.R., Nitao, J.K. and Zangerl, A.R. (1991) Adaptive significance of furanocoumarin diversity in *Pastinaca sativa* (Apiaceae). *Journal of Chemical Ecology* 17(1), 207–215. DOI: 10.1007/BF00994434.

Brodie, B.B., Axelrod, J., Cooper, J.R., Gaudette, L., La Du, B.N. *et al.* (1955) Detoxication of drugs and other foreign compounds by liver microsomes. *Science* 121(3147), 603–604. DOI: 10.1126/science.121.3147.603.

Calla, B., Noble, K., Johnson, R.M., Walden, K.K.O., Schuler, M.A. *et al.* (2017) Cytochrome P450 diversification and hostplant utilization patterns in specialist and generalist moths: birth, death and adaptation. *Molecular Ecology* 26(21), 6021–6035. DOI: 10.1111/mec.14348.

Chen, J., Li, C. and Yang, Z. (2015) Identification and expression of two novel cytochrome P450 genes, CYP6CV1 and CYP9A38, in *Cnaphalocrocis medinalis* (Lepidoptera: Pyralidae). *Journal of Insect Science* 15(1), 50–57. DOI: 10.1093/jisesa/ieu174.

Clarke, J., Wu, H.-C., Jayasinghe, L., Patel, A., Reid, S. *et al.* (2009) Continuous base identification for single-molecule nanopore DNA sequencing. *Nature Nanotechnology* 4(4), 265–270. DOI: 10.1038/nnano.2009.12.

Claudianos, C., Ranson, H., Johnson, R.M., Biswas, S., Schuler, M.A. *et al.* (2006) A deficit of detoxification enzymes: pesticide sensitivity and environmental response in the honeybee. *Insect Molecular Biology* 15(5), 615–636. DOI: 10.1111/j.1365-2583.2006.00672.x.

Cohen, M.B., Schuler, M.A. and Berenbaum, M.R. (1992) A host-inducible cytochrome P-450 from a host-specific caterpillar: molecular cloning and evolution. *Proceedings of the National Academy of Sciences* 89(22), 10920–10924. DOI: 10.1073/pnas.89.22.10920.

D'alençon, E., Sezutsu, H., Legeai, F., Permal, E., Bernard-Samain, S. *et al.* (2010) Extensive synteny conservation of holocentric chromosomes in Lepidoptera despite high rates of local genome rearrangements. *Proceedings of the National Academy of Sciences* 107(17), 7680–7685. DOI: 10.1073/pnas.0910413107.

David, J.P., Strode, C., Vontas, J., Nikou, D., Vaughan, A. *et al.* (2005) The *Anopheles gambiae* detoxification CHIP: a highly specific microarray to study metabolic-based insecticide resistance in malaria vectors. *Proceedings of the National Academy of Sciences* 102(11), 4080–4084. DOI: 10.1073/pnas.0409348102.

Derisi, J., Penland, L., Brown, P.O., Bittner, M.L., Meltzer, P.S. *et al.* (1996) Use of a cDNA microarray to analyse gene expression patterns in human cancer. *Nature Genetics* 14, 457–460.

Derks, M.F.L., Smit, S., Salis, L., Schijlen, E., Bossers, A. *et al.* (2015) The genome of winter moth (Operophtera brumata) provides a genomic perspective on sexual dimorphism and phenology. *Genome Biology and Evolution* 7(8), 2321–2332. DOI: 10.1093/gbe/evv145.

Eid, J., Fehr, A., Gray, J., Luong, K., Lyle, J. *et al.* (2009) Real-Time DNA sequencing from single polymerase molecules. *Science* 323(5910), 133–138. DOI: 10.1126/science.1162986.

Feyereisen, R., Koener, J.F., Farnsworth, D.E. and Nebert, D.W. (1989) Isolation and sequence of cDNA encoding a cytochrome P-450 from an insecticide-resistant strain of the house fly, *Musca domestica*. *Proceedings of the National Academy of Sciences* 86(5), 1465–1469. DOI: 10.1073/pnas.86.5.1465.

Feyereisen, R. (2011) Insect CYP genes and P450 enzymes. In: Gilbert, L.I. (ed.) *Insect Molecular Biology and Biochemistry*. Elsevier, Oxford, UK.

Fodor, S.P., Read, J.L., Pirrung, M.C., Stryer, L., Lu, A.T. *et al.* (1991) Light-directed, spatially addressable parallel chemical synthesis. *Science* 251(4995), 767–773. DOI: 10.1126/science.1990438.

Fujii-Kuriyama, Y., Mizukami, Y., Kawajiri, K., Sogawa, K. and Muramatsu, M. (1982) Primary structure of a cytochrome P-450: coding nucleotide sequence of phenobarbital-inducible cytochrome P-450 cDNA from rat liver. *Proceedings of the National Academy of Sciences* 79(9), 2793–2797. DOI: 10.1073/pnas.79.9.2793.

Gandhi, R., Varak, E. and Goldberg, M.L. (1992) Molecular analysis of a cytochrome P450 gene of family 4 on the *Drosophila* X chromosome. *DNA and Cell Biology* 11(5), 397–404. DOI: 10.1089/dna.1992.11.397.

Gao, S., Ren, Y., Sun, Y., Wu, Z., Ruan, J. *et al.* (2016) Pacbio full-length transcriptome profiling of insect mitochondrial gene expression. *RNA Biology* 13(9), 820–825. DOI: 10.1080/15476286.2016.1197481.

Giraudo, M., Hilliou, F., Fricaux, T., Audant, P., Feyereisen, R. *et al.* (2015) Cytochrome P450s from the fall armyworm (*Spodoptera frugiperda*): responses to plant allelochemicals and pesticides. *Insect Molecular Biology* 24(1), 115–128. DOI: 10.1111/imb.12140.

Gregory, R., Darby, A.C., Irving, H., Coulibaly, M.B., Hughes, M. *et al.* (2011) A *de novo* expression profiling of *Anopheles funestus*, malaria vector in Africa, using 454 pyrosequencing. *PLoS ONE* 6(2), e17418. DOI: 10.1371/journal.pone.0017418.

Guengerich, F.P. (2001) Common and uncommon cytochrome P450 reactions related to metabolism and chemical toxicity. *Chemical Research in Toxicology* 14(6), 611–650. DOI: 10.1021/tx0002583.

International Silkworm Genome Consortium (2008) The genome of a lepidopteran model insect, the silkworm *Bombyx mori*. *Insect Biochemistry and Molecular Biology* 38(12), 1036–1045. DOI: 10.1016/j.ibmb.2008.11.004.

Johnson, R.M., Evans, J.D., Robinson, G.E. and Berenbaum, M.R. (2009) Changes in transcript abundance relating to colony collapse disorder in honey bees (*Apis mellifera*). *Proceedings of the National Academy of Sciences* 106(35), 14790–14795. DOI: 10.1073/pnas.0906970106.

Johnson, R.M., Harpur, B.A., Dogantzis, K.A., Zayed, A. and Berenbaum, M.R. (2018) Genomic footprint of evolution of eusociality in bees: floral food use and CYPome "blooms". *Insectes Sociaux* 65(3), 445–454. DOI: 10.1007/s00040-018-0631-x.

Korf, I. (2013) Genomics: the state of the art in RNA-seq analysis. *Nature Methods* 10(12), 1165–1166. DOI: 10.1038/nmeth.2735.

Krempl, C., Heidel-Fischer, H.M., Jiménez-Alemán, G.H., Reichelt, M., Menezes, R.C. *et al.* (2016) Gossypol toxicity and detoxification in *Helicoverpa armigera* and *Heliothis virescens*. *Insect Biochemistry and Molecular Biology* 78, 69–77. DOI: 10.1016/j.ibmb.2016.09.003.

Le Goff, G., Boundy, S., Daborn, P.J., Yen, J.L., Sofer, L. *et al.* (2003) Microarray analysis of cytochrome P450 mediated insecticide resistance in Drosophila. *Insect Biochemistry and Molecular Biology* 33(7), 701–708. DOI: 10.1016/S0965-1748(03)00064-X.

Lee, S.H., Kang, J.S., Min, J.S., Yoon, K.S., Strycharz, J.P. *et al.* (2010) Decreased detoxification genes and genome size make the human body louse an efficient model to study xenobiotic metabolism. *Insect Molecular Biology* 19(5), 599–615. DOI: 10.1111/j.1365-2583.2010.01024.x.

Levene, M.J., Korlach, J., Turner, S.W., Foquet, M., Craighead, H.G. (2003) Zero-mode waveguides for single-molecule analysis at high concentrations. *Science* 299(5607), 682–686. DOI: 10.1126/science.1079700.

Li, X., Baudry, J., Berenbaum, M.R. and Schuler, M.A. (2004) Structural and functional divergence of insect CYP6B proteins: from specialist to generalist cytochrome P450. *Proceedings of the National Academy of Sciences* 101(9), 2939–2944. DOI: 10.1073/pnas.0308691101.

Lister, R., O'Malley, R.C., Tonti-Filippini, J., Gregory, B.D., Berry, C.C. *et al.* (2008) Highly integrated single-base resolution maps of the epigenome in Arabidopsis. *Cell* 133(3), 523–536. DOI: 10.1016/j.cell.2008.03.029.

Liu, X., Chen, J. and Yang, Z. (2010) Characterization and induction of two cytochrome P450 genes, CYP6AE28 and CYP6AE30, in *Cnaphalocrocis medinalis*: possible involvement in metabolism of rice allelochemicals. *Zeitschrift für Naturforschung C* 65(11-12), 719–725. DOI: 10.1515/znc-2010-11-1213.

Mao, Y.-B., Cai, W.-J., Wang, J.-W., Hong, G.-J., Tao, X.-Y. *et al.* (2007) Silencing a cotton bollworm P450 monooxygenase gene by plant-mediated RNAi impairs larval tolerance of gossypol. *Nature Biotechnology* 25(11), 1307–1313. DOI: 10.1038/nbt1352.

Mao, W., Rupasinghe, S.G., Johnson, R.M., Zangerl, A.R., Schuler, M.A. *et al.* (2009) Quercetin-metabolizing CYP6AS enzymes of the pollinator *Apis mellifera* (Hymenoptera: Apidae). *Comparative Biochemistry and Physiology Part B: Biochemistry and Molecular Biology* 154(4), 427–434. DOI: 10.1016/j.cbpb.2009.08.008.

McGraw, L.A., Gibson, G., Clark, A.G. and Wolfner, M.F. (2004) Genes regulated by mating, sperm, or seminal proteins in mated female *Drosophila melanogaster*. *Current Biology* 14(16), 1509–1514. DOI: 10.1016/j.cub.2004.08.028.

Mortazavi, A., Williams, B.A., Mccue, K., Schaeffer, L. and Wold, B. (2008) Mapping and quantifying mammalian transcriptomes by RNA-seq. *Nature Methods* 5(7), 621–628. DOI: 10.1038/nmeth.1226.

Nagalakshmi, U., Wang, Z., Waern, K., Shou, C., Raha, D. *et al.* (2008) The transcriptional landscape of the yeast genome defined by RNA sequencing. *Science* 320(5881), 1344–1349. DOI: 10.1126/science.1158441.

Nebert, D.W., Adesnik, M., Coon, M.J., Estabrook, R.W., Gonzalez, F.J. *et al.* (1987) The P450 gene superfamily: recommended nomenclature. *DNA* 6(1), 1–11. DOI: 10.1089/dna.1987.6.1.

Nelson, D.R., Kamataki, T., Waxman, D.J., Guengerich, F.P., Estabrook, R.W. *et al.* (1993) The P450 superfamily: update on new sequences, gene mapping, accession numbers, early trivial names of enzymes, and nomenclature. *DNA and Cell Biology* 12(1), 1–51. DOI: 10.1089/dna.1993.12.1.

Nelson, D.R. (1998) Cytochrome P450 nomenclature. In: Phillips, I.R. and Shephard, E.A. (eds) *Cytochrome P450 Protocols*. 107. Humana Press, Totowa, New Jersey.

Nelson, D.R. (2006) Cytochrome P450 nomenclature. In: Phillips, I.R. and Shephard, E.A. (eds) *Cytochrome P450 Protocols*. 320. Humana Press, Totowa, New Jersey.

Nelson, D.R. (2009) The cytochrome P450 Homepage. *Human genomics* 4(1), 59.

Oakeshott, J.G., Johnson, R.M., Berenbaum, M.R., Ranson, H., Cristino, A.S. *et al.* (2010) Metabolic enzymes associated with xenobiotic and chemosensory responses in *Nasonia vitripennis*. *Insect Molecular Biology* 19, 147–163. DOI: 10.1111/j.1365-2583.2009.00961.x.

Omura, T. and Sato, R. (1964) The carbon monoxide-binding pigment of liver microsomes: Evidence for its hemoprotein nature. *Journal of Biological Chemistry* 239, 2370–2378.

Omura, T. (2013) Contribution of cytochrome P450 to the diversification of eukaryotic organisms. *Biotechnology and Applied Biochemistry* 60(1), 4–8. DOI: 10.1002/bab.1099.

Pottier, M.-A., Bozzolan, F., Chertemps, T., Jacquin-Joly, E., Lalouette, L. *et al.* (2012) Cytochrome P450s and cytochrome P450 reductase in the olfactory organ of the cotton leafworm *Spodoptera littoralis*. *Insect Molecular Biology* 21(6), 568–580. DOI: 10.1111/j.1365-2583.2012.01160.x.

Preissner, S., Kroll, K., Dunkel, M., Senger, C., Goldsobel, G. *et al.* (2010) SuperCYP: a comprehensive database on cytochrome P450 enzymes including a tool for analysis of CYP-drug interactions. *Nucleic Acids Research* 38(suppl_1), D237–D243. DOI: 10.1093/nar/gkp970.

Ranson, H., Claudianos, C., Ortelli, F., Abgrall, C., Hemingway, J. *et al.* (2002) Evolution of supergene families associated with insecticide resistance. *Science* 298(5591), 179–181. DOI: 10.1126/science.1076781.

Richards, S., Gibbs, R.A., Weinstock, G.M., Brown, S.J., Denell, R. *et al.* (2008) The genome of the model beetle and pest *Tribolium castaneum*. *Nature* 452(7190), 949.

Rose, R.L., Barbhaiya, L., Roe, R.M., Rock, G.C. and Hodgson, E. (1995) Cytochrome P450-associated insecticide resistance and the development of biochemical diagnostic assays in *Heliothis virescens*. *Pesticide Biochemistry and Physiology* 51(3), 178–191. DOI: 10.1006/pest.1995.1018.

Sirim, D., Wagner, F., Lisitsa, A. and Pleiss, J. (2009) The cytochrome P450 engineering database: integration of biochemical properties. *BMC Biochemistry* 10(1), 27. DOI: 10.1186/1471-2091-10-27.

Stevens, J.L., Snyder, M.J., Koener, J.F. and Feyereisen, R. (2000) Inducible P450s of the CYP9 family from larval *Manduca sexta* midgut. *Insect Biochemistry and Molecular Biology* 30(7), 559–568. DOI: 10.1016/S0965-1748(00)00024-2.

Tijet, N., Helvig, C. and Feyereisen, R. (2001) The cytochrome P450 gene superfamily in *Drosophila melanogaster*: annotation, intron-exon organization and phylogeny. *Gene* 262(1-2), 189–198. DOI: 10.1016/S0378-1119(00)00533-3.

Wang, B., Shahzad, M.F., Zhang, Z., Sun, H., Han, P. *et al.* (2014) Genome-wide analysis reveals the expansion of cytochrome P450 genes associated with xenobiotic metabolism in rice striped stem borer, *Chilo suppressalis*. *Biochemical and Biophysical Research Communications* 443(2), 756–760. DOI: 10.1016/j.bbrc.2013.12.045.

Wang, H., Shi, Y., Wang, L., Liu, S., Wu, S. *et al.* (2018) CYP6AE gene cluster knockout in *Helicoverpa armigera* reveals role in detoxification of phytochemicals and insecticides. *Nature Communications* 9(1), 4820. DOI: 10.1038/s41467-018-07226-6.

Wang, R.-L., Staehelin, C., Xia, Q.-Q., Su, Y.-J. and Zeng, R.-S. (2015) Identification and characterization of CYP9A40 from the tobacco cutworm moth (*Spodoptera litura*), a cytochrome P450 gene induced by plant allelochemicals and insecticides. *International Journal of Molecular Sciences* 16(9), 22606–22620. DOI: 10.3390/ijms160922606.

Yin, C., Shen, G., Guo, D., Wang, S., Ma, X. *et al.* (2016) InsectBase: a resource for insect genomes and transcriptomes. *Nucleic Acids Research* 44(D1), D801–D807. DOI: 10.1093/nar/gkv1204.

Yu, L., Tang, W., He, W., Ma, X., Vasseur, L. *et al.* (2015) Characterization and expression of the cytochrome P450 gene family in diamondback moth, *Plutella xylostella* (L.). *Scientific Reports* 5(1), 8952. DOI: 10.1038/srep08952.

Zagrobelny, M., Scheibye-Alsing, K., Jensen, N.B., Møller, B.L., Gorodkin, J. *et al.* (2009) 454 pyrosequencing based transcriptome analysis of *Zygaena filipendulae* with focus on genes involved in biosynthesis of cyanogenic glucosides. *BMC Genomics* 10(1), 574. DOI: 10.1186/1471-2164-10-574.

7 Whole-Body Transcriptome of the Douglas-Fir Seed Chalcid, *Megastigmus spermotrophus*, Reveals Ecological and Evolutionary Insights

Amber R. Paulson,* Jürgen Ehlting, Patrick von Aderkas, and Steven J. Perlman

Department of Biology, University of Victoria, British Columbia, Canada

7.1 Introduction

Within the astonishingly diverse and successful Hymenoptera, parasitoids represent the largest and most speciose lineage. All parasitoids are characterized by having a free-living adult stage and a juvenile stage that develops on or within an animal host (usually another insect), ultimately killing it (Eggleton and Belshaw, 1992). Parasitoids have evolved to attack a wide range of hosts, from spider egg cases to aquatic insects to butterfly eggs (LaSalle and Gauld, 1991). Successful development of parasitoid offspring within or upon a host relies on a cocktail of venoms and other host-modulating effectors. In addition to venom, some major endoparasitoid lineages have domesticated large, symbiotic polydnaviruses or produce virus-like particles that deliver virulence factors into the parasitized host (Herniou *et al.*, 2013; Drezen *et al.*, 2017). The study of parasitoid wasp venoms and symbiotic viruses has been given a boost recently by transcriptomic investigations, which provide insight into the evolution and ecology of such fascinating, yet often cryptic systems. So, the aim of this chapter is to show the utility of whole-body transcriptomics in the study of wasp venom evolution and of associated microorganisms, such as viruses, bacteria and intracellular parasites.

We will focus on a unique chalcid wasp, *Megastigmus spermotrophus* Wachtl, that has secondarily evolved to parasitize conifer trees using unknown mechanisms of host manipulation. Among parasitoid wasps, one

*Corresponding author: amber.rose.paulson@gmail.com

can find several independent examples of extreme host transitions, in which species have evolved from animal parasitism to plant parasitism by adopting an endophytophagous lifestyle. For example, secondary evolution of galling occurs in the Braconidae, Cynipidae and Chalcidoidea; and secondary evolution of seed-feeding occurs in the Chalcidoidea (Austin and Dangerfield, 1998; Ronquist and Liljeblad, 2001; Munro *et al.*, 2011). Seed-feeding chalcids in the genus *Megastigmus* represent a fascinating and successful group of invasive plant parasites. There are currently 134 species of *Megastigmus*, of which 72 are seed feeders (Grissell, 1999; Auger-Rozenberg and Roques, 2012). Of the remaining species, the majority are gall-inducers, inquilines or facultative parasitoids of other gall-inducing wasps (Grissell, 1999).

Seed-feeding species of *Megastigmus* are well described invasive plant pests. In European surveys of this genus, 11 of 21 species of seed-feeding *Megastigmus* were found to be of exotic origin (Roques and Skrzypczyńska, 2003; Auger-Rozenberg *et al.*, 2006). Most introduced *Megastigmus* spp. found in Europe were originally associated with conifers native to the Nearctic (Roques and Skrzypczyńska, 2003). A common source of accidentally introduced seed-feeding pests is exotic seed material that has been inadequately inspected (Roques and Skrzypczyńska, 2003). Infestations by populations of invasive *Megastigmus* spp. not only are a problem for commercially important tree species, but also, depending on the host range of the insect, may threaten other native tree species. Some *Megastigmus* are narrow in their host range, i.e. they are species-specific, while others exhibit genus-level host specificity. Both *Megastigmus pinus* Parfitt and *M. rafni* Hoffmeyer can now be found on the European silver fir, *Abies alba*, in southern France (Roques and Skrzypczyńska, 2003; Auger-Rozenberg *et al.*, 2006). In addition, *Megastigmus schimitscheki* Novitzky switched from the Cypriot endemic conifer *Cedrus brevifolia* to *Cedrus atlantica* when the insect was introduced into France (Auger-Rozenberg *et al.*, 2012). Infestation of natural stands may have a negative effect on forest regeneration and age-demographic structure over the long term. The introduction of *Megastigmus* could also increase competition for seed resources with native seed-feeding animals in introduced areas (Fabre *et al.*, 2004).

Many species of conifer-associated *Megastigmus* possess biological attributes that contribute to their invasive potential (Roques *et al.*, 2003). Most species of conifer-associated *Megastigmus* have evolved the ability to remain viable for up to 5 years in a state of extended larval diapause, an adaptive strategy that allows them to survive annual fluctuations in host-tree seed production (Turgeon *et al.*, 1994; Roques *et al.*, 2003). Because of diapause, larvae are able to survive not only well past seed harvest, but also through processing, shipment, and prolonged seed storage at the new destination. Another adaptation is parthenogenetic reproduction, which increases the chance of successful establishment in an introduced area. In addition, some *Megastigmus* spp. reproduce asexually following infection by a vertically transmitted bacterium, *Wolbachia* (Boivin and Candau, 2007; Boivin *et al.*, 2008, 2014).

The best-studied species in the genus is *M. spermotrophus*. Its host is Douglas-fir, *Pseudotsuga menziesii*. By manipulating its host, the insect induces the steady accumulation of nutrient reserves within the host seeds' storage tissues throughout larval development (von Aderkas *et al.*, 2005a). Remarkably, even if insect eggs are laid into Douglas-fir ovules, i.e. unfertilized megagametophtyes, the parasite is able to redirect development of the megagametophytes such that accumulation of storage products continues to the sole benefit of its developing larvae (von Aderkas *et al.*, 2005b). By manipulating seed development in this way, *M. spermotrophus* co-opts the conifer female reproductive tissue for its own reproductive success (von Aderkas *et al.*, 2005a, b). The exact mechanism utilized by *M. spermotrophus* to alter Douglas-fir seed development is not currently known, although data from hormone profiling suggested that manipulation by the insect of a particular class of plant hormones, i.e. cytokinins, may contribute to effective seed parasitism (Chiwocha *et al.*, 2007).

This chapter will describe the application of whole-body transcriptomics as a method to investigate a group of insects that are emerging as a globally invasive plant parasite. In this context, transcriptomics is a powerful tool that can generate valuable and novel information about pest insects, especially those that are difficult to maintain in standard laboratory settings, which is the case for *M. spermotrophus*. Key findings from this work related to the *de novo* assembly and annotation of the *M. spermotrophus* transcriptome have provided insights into venom evolution, e.g. identification of aspartylglucosaminidase, a putative venom component that likely arose through gene duplication (Paulson *et al.*, 2016). Additional findings from the transcriptome, including identification of a novel microsporidium and positive-sense single-stranded RNA ((+)ssRNA) virus, will be further detailed here.

7.2 Materials and Methods

7.2.1 RNA extraction

Laboratory-reared adults (males and females) and diapaused larvae of *M. spermotrophus* were obtained from heavily infested seed from the Mt Newton Seed Orchard, located in Saanichton, British Columbia (BC) (48°35′54.00"N, 123°25′56.87"W). Wild females were collected from trees located on the University of Victoria campus in Victoria, BC (48°27′42.90"N, 123°18′37.50"W). Approximately 10–20 individuals were placed into 2 ml Micro tubes (Starstedt) with 700 µl buffer RLT (Qiagen), 1/100 volume beta-mercaptoethanol and three 3.5 mm diameter glass beads (BioSpec Products). Samples were homogenized using the Mini-Beadbeater (BioSpec Products) at half-speed for 90 s. The homogenate was centrifuged at 1300 × *g* for 3 min. Total RNA was extracted using RNeasy (Qiagen), followed by on-column DNase digestion and RNA clean-up, using the manufacturer's guidelines. Precipitated RNA was analyzed using an Experion Electrophoresis Station (Bio-Rad). Complementary DNA library construction was done by using oligo(dT) primers, followed by library fragmentation, size exclusion purifications (target average sequence length of 300bp) and sequencing on the Illumina sequencing platform (HISeq 2000) at the BC Cancer Agency Genome Sciences Centre, Vancouver, Canada.

7.2.2 Short-read filtering and *de novo* assembly

Short reads were first filtered with Trimmomatic (v0.22) (Bolger *et al.*, 2014) with the following parameters: minimum leading quality of three, minimum trailing quality of 20, minimum read length of 36 and sliding window of four bases with a minimum quality of 20. These were assembled *de novo* using the trans-ABySS (v1.3.2) pipeline; we used k-mer values of 30, 35, and even values from 52 to 96 (Robertson *et al.*, 2010). The assembly was further clustered using CD-HIT-EST (v4.6) with the default sequence identity threshold of 0.95 (Li and Godzik, 2006). Additional clustering was performed by using TIGR-TGICL (Pertea *et al.*, 2003) with the Cap3 specific overlap percentage identity cut-off set to 98. Only transcripts larger than 100 bases were used in subsequent analysis. *M. spermotrophus* short-reads libraries and final *de novo* transcriptome assembly have been deposited with GenBank under the accession GCPB00000000.1.

7.2.3 Annotation

The genome of parasitoid chalcid *Nasonia vitripennis* (an ectoparasitoid of flesh fly pupae) provided an important resource to annotate the *de novo* transcriptome of *M. spermotrophus* (Werren *et al.*, 2010). Supplemental to the genome, a recent study identified 79 constituents of *Nasonia* venom, obtained by a combination of bioinformatics and proteomics (de Graaf *et al.*, 2010). In order to identify putative venom transcript homologs, the *N. vitripennis* venom protein query was compared with the *M. spermotrophus* transcriptome using tBLASTn, with an E-value cut-off of 1×10^{-15}. In addition to the venom annotations, the *M. spermotrophus* transcriptome was annotated with BLASTX (v2.2.27+) against GenBank's non-redundant (nr) database (as of January 2013) with an E-value cut-off of 1×10^{-5} (Camacho *et al.*, 2009). Any transcripts without BLASTx hits were then annotated with BLASTn using GenBank's nucleotide database (as of January 2013) with the same E-value cut-off. Corresponding taxonomic information was linked to each annotation by the taxon identifier from GenBank's taxonomy database.

7.2.4 Differential expression analysis

Transcript expression was quantified using RSEM (v1.2.0) by aligning forward reads only and by providing a mean fragment length of 300bp

(Li and Dewey, 2011). Only transcripts with NCBI annotations were considered for expression analysis. Three normalization methods (trimmed mean of M (TMM), upper quartile (UQN) and conditional quantile normalization (CQN)) were evaluated by comparing count distribution variability among libraries. The data were finally normalized using the CQN R package (v1.7.0) (Hansen *et al.*, 2012).

Differential expression analysis was implemented using the nonparametric statistical analysis package Noiseq (v2.0.0) (Tarazona *et al.*, 2011). The Noiseq-sim feature was utilized to simulate technical replicates with the following parameters: the size of simulated samples was set as equal to 20% of sequencing depth, five simulation replicates were done with an allowance of 2% variability. Bioinformatics packages were implemented using R (v3.0.1) (R Core Team, 2014) in RStudio (v0.97.551) (RStudio Team, 2014).

7.2.5 Microsporidia validation

Samples of wild adult female *M. spermotrophus* were collected as above and placed individually into 2 ml Micro tubes (Sarstedt) with 100 µl of PrepMan Ultra Reagent (Applied Biosystems, USA) that contained 20 zirconia or silica beads of 1 mm in diameter (BioSpec Products). Genomic DNA was extracted from the samples by homogenization using the Mini-Beadbeater-16 (BioSpec Products) on maximum (3450 oscillations/min) for two 20–30 s cycles separated by 30 s of centrifugation at 13,000 × *g*. The samples were then incubated at 100°C for 10 min, then room temperature for 1 min, then centrifuged for 3 min at 13,000 × *g* and the supernatant transferred into new Eppendorf tubes.

Samples were screened for the presence of microsporidia using 18S ribosomal RNA primers: 18-MicroF (5′-CACCAGGTTGATTCTGCC-3′) (Baker *et al.*, 1995) and NemopopR (5′-CGGTACAACGGTCTCTA-3′) (Alex Ardila-Garcia and Naomi Fast, Vancouver, 2012, personal communication) using ABM PCR Taq and reagents. PCR reactions were conducted using the following temperature profile: 94°C for 3 min, followed by 30 cycles of 94°C for 1 min, 53.7°C for 1 min and 72°C for 1.5 min, with a final elongation step at 72°C for 5 min. The quality of the template DNA was verified by PCR amplification of the CO1 gene with primers LCO1490 (5′-GGTCAACAAATC ATAAAGATATTGG-3′) and HCO2198 (5′-TAAACTTCAGGGTGACCAAA AAATCA-3′) (Folmer *et al.*, 1994).

To obtain a longer microsporidian 18S rRNA sequence for phylogenetic analysis, amplicon products from 18-MicroF and 1537 r (5′-TTATGATCCTGC TAATGGTTC-3′) (Baker *et al.*, 1995) were cloned using the Strataclone kit with Solopack Competent cells (Stratagene). Sequencing was performed by Macrogen USA using standard M13F/R primers and a consensus 18S rRNA from this microsporidium was submitted to GenBank under accession number MH712042.

7.2.6 Phylogenetic analysis

Phylogenetic assignment of the microsporidium identified in the *M. spermotrophus* transcriptome was completed by evolutionary reconstruction using 18S rRNA sequences from a broad range of Clade IV Microsporidia (Baker *et al.*, 1995). Sequences were initially aligned using ClustalW, then visually inspected and trimmed as required. A maximum-likelihood tree was generated using the best predicted DNA model (Tamura-Nei model and gamma distributed rates among sites model of nucleotide substitution) with MEGA (v7.0.18) (Tamura *et al.*, 2011). Branch support was assessed by bootstrapping the data 1000 times.

An entire genome from a (+)ssRNA virus was deposited into GenBank under Accession No. MH737698. Viral open reading frames (ORFs) were determined using the 'Gene Finding in Viral Genomes' algorithm in Softberry (http://linux1.softberry.com/all.htm). Putative conserved protein domains were assigned using BLAST against the NR protein database (as of June, 2018). Additional putative features were assigned using predicted secondary structure features generated by Phyre2 (Kelley *et al.*, 2015). A phylogenetic reconstruction was derived using RdRp amino acid sequence from the (+)ssRNA virus genome and a selection of RdRp sequences from the 'Hepe-Virga' clade of (+)ssRNA viruses (Shi *et al.*, 2016). The phylogeny was made using MEGA with the best predicted protein model (LG model with frequencies and discrete gamma distribution with invariant sites for evolutionary rate differences) and branch support was determined by bootstrapping the data 500 times.

7.3 Results

7.3.1 Short-read filtering and *de novo* assembly

The *M. spermotrophus* whole-insect cDNA libraries (larva, adult male, laboratory-reared adult female, and wild adult female) consisted of 47.40 Gbps of 100 bp paired-end sequence data, with fewer reads sequenced from the wild adult female (Table 7.1). The transcriptome of *M. spermotrophus* was assembled *de novo* using multiple k-mer values, followed by combining each individual assembly using trans-ABySS. This resulted in 1,361,656 assembled transcripts (Table 7.2). Clustering by CD-HIT-EST and TIGR-TGICL resulted in 44,176 clusters and 149,236 singletons. The final set of 143,306 transcripts ranged in length from 101 (minimum transcript length) up to 32,049 bp, with N50 of 2420 bp.

7.3.2 Annotation

Transcripts from the *M. spermotrophus* transcriptome were annotated using protein and nucleotide sequence homology searches against GenBank's NR

Table 7.1. Illumina sequencing output for the *Megastigmus spermotrophus* whole-body cDNA.

Library	Larva	Male	Lab-reared female	Wild female	Total
Total number raw paired-end reads	62,396,989	64,991,169	63,560,118	46,037,319	**236,985,595**
Raw paired-end read total length (Gbp)	12.48	13.00	12.71	9.21	**47.40**
Total number filtered: paired-end reads	49,883,162	52,200,155	50,592,674	36,311,907	**188,987,898**
Total number filtered: single-end reads	6,023,891	6,299,194	6,135,548	4,478,658	**22,937,291**

Table 7.2. *Megastigmus spermotrophus* transcriptome clustering results.

Clustering method	Number of transcripts	N50 (min. 200 bp)
Trans-ABySS	1,361,656	1,690
CD-HIT-EST	296,711	1,570
TIGR-TGICL	193,412	2,420
Singleton removal	143,306	2,420

protein and nucleotide databases using an E-value threshold of $1e10^{-5}$. Only 27.6 and 2.1% of the transcripts were assigned annotations from the protein and nucleotide databases, respectively. Many of the annotations were redundant and were assigned to multiple transcripts, though they possibly represent transcript variants. In total, 14.2% of transcripts were assigned non-redundant annotations (Table 7.3). Many of the annotations were assigned as 'hypothetical' or 'predicted' proteins that do not currently have a known function. In addition, most small transcripts (less than 600 bp) were not assigned an annotation (data not shown). A total of 21 putative venom annotations were assigned to the *M. spermotrophus* transcriptome using the *N. vitripennis* venom database with a more stringent E-value threshold of 1×10^{-15} (Paulson *et al.*, 2016).

Taxon identifiers associated with each non-redundant annotation were used to correlate the taxonomic source of the most homologous sequences in the protein database. As expected, the majority of annotations were assigned from insects (77.3%) (Table 7.4). *N. vitripennis* was the source of most insect annotations (61.2%) (Table 7.5). A variety of Hymenoptera, including *Apis* spp., *Bombus* spp., *Camponotus floridanus, Megachile rotundata, Harpegnathos saltator, Solenopsis invicta* and *Acromyrmex echinatior*, were also well represented in the transcriptome data.

Annotations from prokaryotes were also identified in the transcriptome (14.8%). *Ralstonia* (Betaproteobacteria) was particularly well-represented: 10.6% of all annotations assigned from the protein database were from this genus (Table 7.4). Another well represented group was the microsporidia, which are intracellular parasites. Microsporidia represented 1.5% of all non-redundant annotations and nearly half (49.5%) of all fungal annotations. Putative viral transcripts represented 0.71% of all non-redundant annotations.

Table 7.3. *Megastigmus spermotrophus* transcriptome annotation results (E-value cut-off = 1×10^{-5}).

Database	NCBI nr protein	NCBI nucleotide[a]	Total
Total annotations	39,600	3,077	**42,677**
Total non-redundant	18,610	1,674	**20,284**

[a]NCBI nucleotide annotations only assigned to transcripts that were not previously assigned an NCBI protein annotation.

Table 7.4. *M. spermotrophus* transcriptome taxonomic annotations based on homology to non-redundant protein database.

Taxonomic group	Proportion (% total)[a]
Animal	81.8 %
Insect (77.3 %)	
Other (4.5 %)	
Bacteria	14.8 %
Ralstonia spp. (10.6 %)	
Other (4.2 %)	
Fungus	1.5 %
Plant	0.8 %
Virus	0.7 %
Protist	0.3 %
Other	0.1 %

[a]Proportions are reported for all non-redundant annotations (shorter transcript variants sharing the same annotation have been removed)

7.3.3 Normalization and differential expression

Expression values were estimated by mapping forward reads on to the transcripts assembly. Only 3.0% of reads failed to align. Expression values were only calculated for transcripts of length 300bp or longer. Expected count data were normalized using CQN, which resulted in better standardization of count distributions across libraries than TMM and UQN (Fig. 7.1).

Table 7.5. *M. spermotrophus* transcriptome insect annotations based on homology to non-redundant protein database.

Insect group	Proportion[a]
Nasonia vitripennis	61.2%
Apis spp.	5.4%
Bombus spp.	5.3%
Camponotus floridanus	5.1%
Megachile rotundata	4.3%
Harpegnathos saltator	3.9%
Solenopsis invicta	3.8%
Acromyrmex echinatior	2.7%
Other insects	2.4%
Tribolium castaneum	2.1%
Drosophila spp.	1.8%
Acyrthosiphon pisum	1.0%
Mosquitos	0.9%

[a]Proportions are reported for all non-redundant annotations (shorter transcript variants sharing the same annotation have been removed).

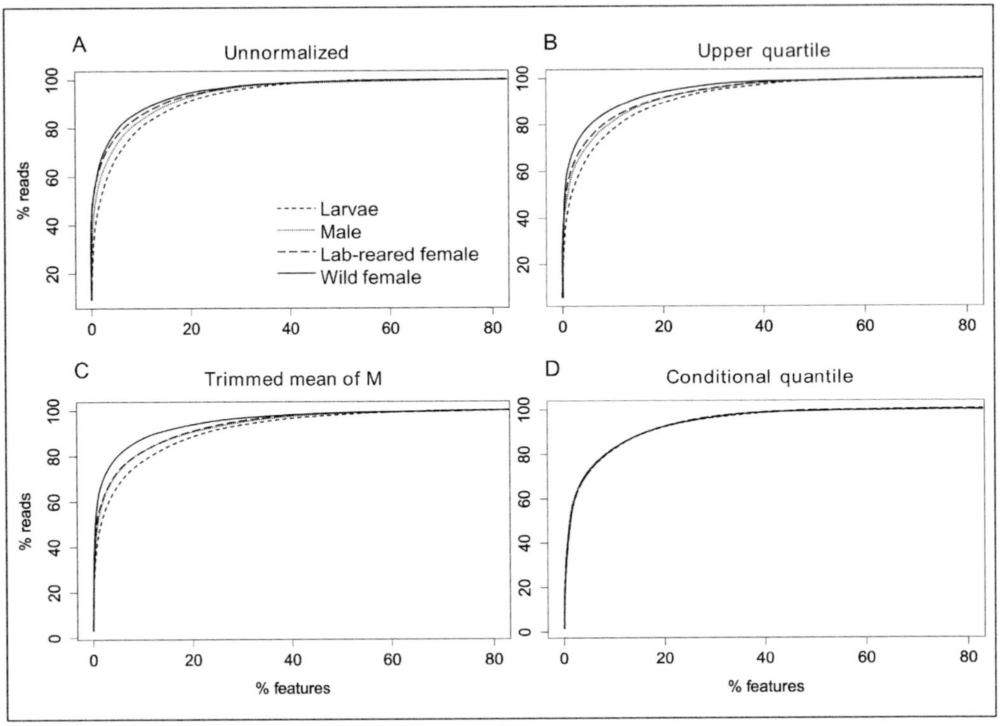

Fig. 7.1. Comparison of count distribution plots of *M. spermotrophus* transcriptome data. (A) Unnormalized. (B) Upper quartile normalization. (C) Trimmed mean of M normalization. (D) Conditional quantile normalization

The NOISeq-sim algorithm identified 518 transcripts that are likely differentially expressed (differential expression probability q-value ≥0.8) in *M. spermotrophus* females compared with larvae and males (data not shown). Three putative venom precursors (aspartylglucosaminidase, Gram-negative bacteria binding 1–2 and venom protein R) were identified as having generally higher expression in females compared with males and larvae (Paulson *et al.*, 2016). Transcripts from a novel intracellular parasitic microsporidium and (+)ssRNA virus (Table 7.6) were also found to have a high probability of differential expression in females compared with males and larvae (Table 7.7).

7.3.4 Microsporidia investigation

Two microsporidian transcripts were found to be highly expressed in both wild female libraries, compared with other libraries (Table 7.7), and further screening of DNA isolated from wild female samples confirmed the presence of microsporidia in a naturally occurring population of *M. spermotrophus* in British Columbia. Using microsporidia-specific primers, a portion of the 18S rRNA gene from this novel microsporidium was amplified and validated

Table 7.6. Annotation details for microsporidia and (+)ssRNA virus transcripts with significantly higher expression in female libraries.

Transcript ID	Length (bp)	Annotation	Taxonomic assignment	Query cover	Identities[a]	E-value
Microsporidia						
Contig99171	1,393	Hypothetical protein NCER_100828	*Nosema ceranae* BRL01	47%	80/222 (36%)	1×10^{-29}
Contig73577	1,069	Hypothetical protein NCER_101591	*Nosema ceranae* BRL01	24%	43/87 (49%)	1×10^{-12}
(+)ssRNA virus						
Contig39309	12,061	207K protein	Soil-born cereal mosaic virus	21%	269/873 (31%)	2×10^{-89}
Contig191610	2,534	Replicase-associated protein	*Solanum violaefolium* ringspot virus	13%	46/110 (42%)	9×10^{-18}

[a]Number of exact amino acid matches between query and subject.

by sequencing from a single wild-captured female *M. spermotrophus* (due to limited sample availability only six individual samples were screened and one individual insect tested positive for microsporidia). To place the microsporidia into phylogenetic context, a maximum likelihood tree was generated based on the 18S rRNA sequence (Fig. 7.2). The unknown microsporidium was allied with the *Nosema/Vairimorpha* subgroup within Clade IV, which is of terrestrial origin (Vossbrinck and Debrunner-Vossbrinck, 2005).

7.3.5 Positive-sense single-stranded RNA virus investigation

A novel (+)ssRNA virus genome of 12 kb length was assembled *de novo* from the *M. spermotrophus* transcriptome, and was found to have greater expression in the females compared with either males or larvae (Table 7.7). Within the viral genome, seven putative ORFs were predicted using fgenesV (Fig. 7.3). Based on significant sequence similarity to conserved protein domains in GenBank's nr protein database, we identified in the virus genome a viral methyltransferase (pfam01660), ftsJ-like methyltransferase (pfam01728), RNA helicase (pfam01443), RNA-dependent RNA polymerase (RdRp) (cd01699), and capsid protein. In addition, a putative hydrolase domain was identified in the first ORF based on similarity to predicted protein secondary structure (Protein Data Bank Code c3rnvA). A 2.5 kb transcript variant with very high sequence homology to the (+)ssRNA virus regions

Table 7.7. Expression details for microsporidia and (+)ssRNA virus transcripts with significantly higher expression in female libraries.

Transcript ID	CQN normalized expression				Mean normalized expression		
	Larvae	Male	Lab-reared female	Wild female	Larvae/ male	female	DE probability[a]
Microsporidia							
Contig99171	0.14	0.28	0.13	87.49	0.21	43.81	0.88
Contig73577	0.17	0.31	0.15	88.34	0.24	44.25	0.88
(+)ssRNA virus							
Contig39309	0.53	0.01	20.70	40.83	0.27	30.77	0.84
Contig191610	1.07	0.97	17.42	146.78	1.02	82.10	0.92

[a]Differential expression probability (*q*-value) using Noiseq-sim comparing female vs non-female normalized expression (at *q*-value 0.8, a transcript is four times more likely to be differentially expressed than not differentially expressed).

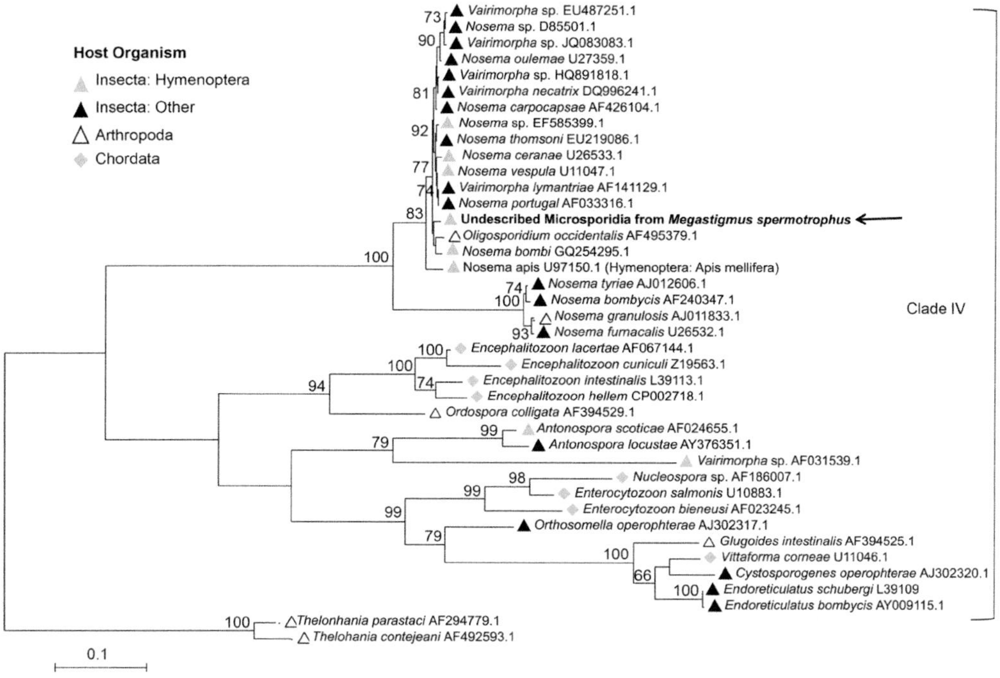

Fig. 7.2. Maximum likelihood phylogeny for microsporidian 18S rRNA sequence using the general time reversible and gamma distributed rates among sites model of nucleotide substitution. Arrow indicating the sequence generated by this study. Numbers next to the nodes indicate percentage of bootstrap support from 1000 bootstrap replicates. Nodes without numbers received less than 65% bootstrap support.

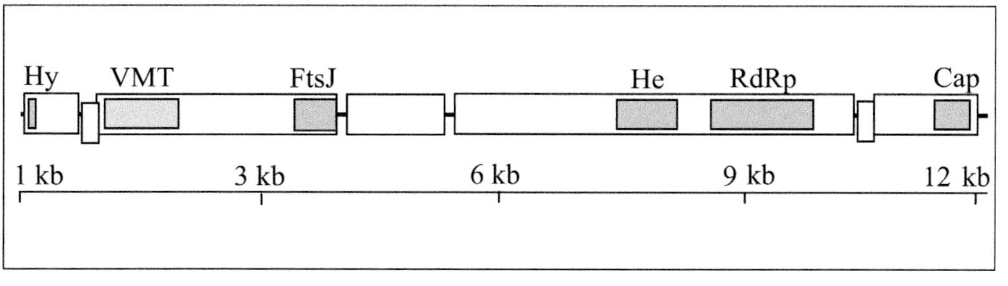

Fig. 7.3. Genomic annotation of the undescribed positive single-stranded RNA virus assembled from the *Megastigmus spermotrophus* transcriptome. Open reading frames (ORFs) (white boxes) were identified using fgenesV. Putative conserved protein domains (gray boxes) were assigned based on sequence homology of translated ORFS against the GenBank non-redundant protein database or secondary protein structure similarities using Phyre2, respectively. Hy, putative hydrolase; VMT, viral methyltransferase; FtsJ, FtsJ-like methyltransferase; He, viral RNA helicase; RdRp, RNA-dependent RNA polymerase; Cap, putative capsid protein.

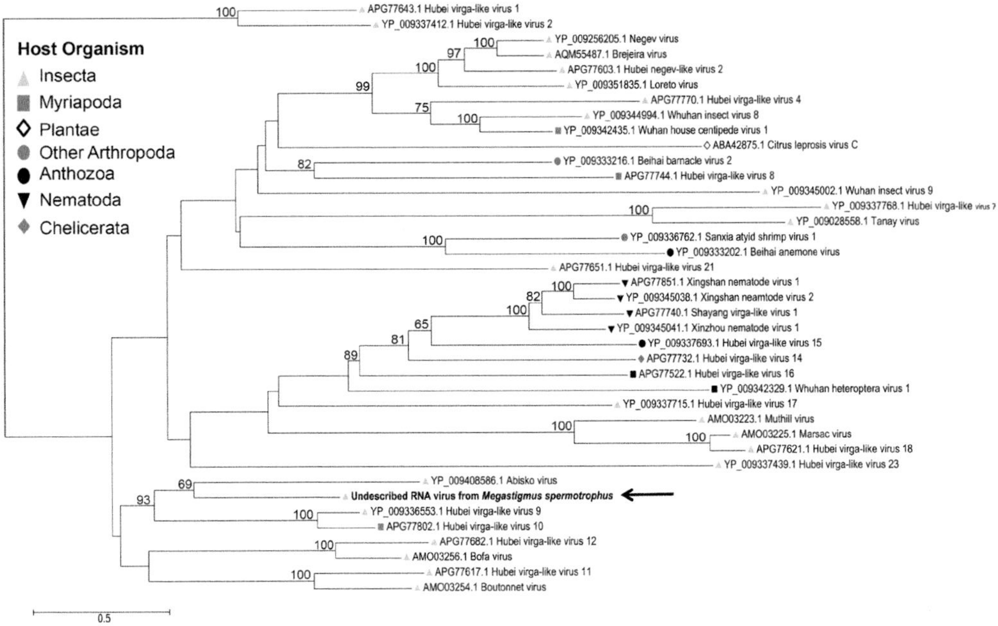

Fig. 7.4. Maximum likelihood phylogeny for (+)ssRNA virus RdRp sequence using the LG model of nucleotide substitution and gamma distributed rates with invariant sites. Arrow indicates the sequence generated by this study. Numbers next to the nodes indicate percentage of bootstrap support from 500 bootstrap replicates. Nodes without numbers received less than 65% bootstrap support.

containing (a small 3′- portion of) the RdRp and capsid protein from the (+) ssRNA virus was also identified as highly expressed in female libraries.

The (+)ssRNA virus identified from the *M. spermotrophus* transcriptome could be placed among the 'Hepe-Virga' clade of (+)ssRNA viruses based on alignment of the RdRp protein (adapted from Shi *et al.*, 2016) (Fig. 7.4). The RdRp protein sequenced from the recently isolated Abisko virus shared the highest homology to the (+)ssRNA virus from *M. spermotrophus*.

7.4 Discussion

7.4.1 General features of the *M. spermotrophus* transcriptome

De novo assembly and post-assembly clustering of the *M. spermotrophus* transcriptome generated 143,306 assembled transcripts with a relatively large average transcript length of over 800bp. Around 70% of transcripts from the *M. spermotrophus* transcriptome were not assigned an annotation at all. Many of these transcripts, frequently of short length, are likely small, non-coding RNAs or partially assembled transcripts that were too short to be assigned using standard sequence homology searches. Over half of the annotations that were assigned were redundant; that is, they were assigned to more than one transcript. Although this redundancy contributed additional noise to the transcriptome data (Yang and Smith, 2013), it remains unclear how much of this was due to the assembly method, assembly errors or genomic DNA contamination, rather than to actual transcriptomic variation in *M. spermotrophus* resulting from recent gene duplications, allelic variation or alternative slice variants.

Of those transcripts assigned an annotation, most were identified from the closely related model parasitoid *N. vitripennis*. The majority (77.3%) of non-redundant annotations were of insect origin. The transcriptome contained a number of genes that are known to be associated with oogenesis, such as vitellogenin and vitellogenin-like protein (Guidugli *et al.*, 2005), vitellogenin receptor (Schonbaum *et al.*, 2000), nanos (Forbes and Lehmann, 1998) and the maternal effect protein oskar (Lehmann and Nüsslein-Volhard, 1986). These genes were found to be more highly expressed in females than in males or larvae. Other interesting transcripts were related to odor perception, such as chemosensory protein CSP-1 (Pelosi *et al.*, 2006) and putative odorant binding protein 70 (Vogt *et al.*, 1999). Again, these were preferentially expressed in females, suggesting that female *M. spermotrophus* utilize chemical cues to locate susceptible host trees.

While the RNA-seq library preparation method targeted eukaryotic messenger RNA containing poly(A) tails, the *M. spermotrophus* transcriptome also included a relatively large number of transcripts of prokaryote origin. We found that 10% of non-redundant annotations were assigned to a single bacterial genus *Ralstonia* (Betaproteobacteria). This result corroborates our previous work in which we performed 16S rRNA amplification of whole-body DNA extracted from *M. spermotrophus*. We found *Ralstonia* to be pervasively associated with *M. spermotrophus* (Paulson *et al.*, 2014). Interestingly, the four most highly expressed *Ralstonia* transcripts are related to transposable elements (TEs), including: transposase IS66 (YP_001899056.1), ISPsy11 transposase OrfB (ZP_10987507.1), resolvase domain protein (ZP_10982688.1) and putative cointegrate resolution protein T (ZP_07678195.1). Expression of TEs by *Ralstonia* within *M. spermotrophus* may be indicative of a symbiont transitioning into a more host-specific, or obligate lifestyle: a large increase in the frequency of TEs is thought to contribute typically to gene loss/gain and rearrangements that are observed within bacterial genomes that have recently experienced selective pressure due to host restriction (Moran and Plague, 2004; Newton and Bordenstein, 2011).

7.4.2 Identification of uncharacterized microsporidia in wild female *M. spermotrophus*

Microsporidia are common intracellular parasites and pathogens of animals, including insects. Recent molecular work has placed the phylum Microsporidia within the kingdom Fungi (Hirt *et al.*, 1999; Capella-Gutiérrez *et al.*, 2012). Here we report an undescribed species of microsporidium that infects *M. spermotrophus*. It was initially identified from whole-body transcriptome libraries of field-collected females. The overall expression of fungal transcripts in the transcriptome was almost without exception exclusive to wild females. Microsporidian infection of wild females was confirmed by targeted PCR screening.

Indirect support for microsporidian infection comes from immunity-related transcripts that were highly expressed in females. These proteins included akirin-like protein (Goto *et al.*, 2008), Gram-negative bacteria binding 1–2 precursor and beta-1,3-glucan binding proteins (Kim *et al.*, 2000), antimicrobial peptide defensin 1–2 precursor (Hoffmann and Hetru, 1992), and hymenoptaecin-1 isoform 2 precursor (Casteels *et al.*, 1993). The differential expression of immunity-related transcripts, especially fungal-specific immunity binding proteins, suggests a host reaction to a pathogen, such as Microsporidia.

Phylogenetic analysis of the 18S small subunit rRNA showed that the *Megastigmus*-associated microsporidium lies in the *Nosema/Varimorpha* clade, which includes important parasites of terrestrial insects (Solter *et al.*, 2012), such as *Nosema apis* and *Nosema ceranae* that infect the honey bee (Fries, 1993; Fries *et al.*, 1996). Microsporidia have also been identified among chalcid parasitoid wasps that target muscoid flies, such as *Tachinaephagus zealandicus*, *Muscidifurax raptor*, *M. zaraptor* and *Urolepis rufipes* (Zchori-Fein *et al.*, 1992; Becnel and Geden, 1994; Geden *et al.*, 2003). In both *T. zealandicus* and *M. raptor*, Microsporidia were vertically transmitted and infected females had reduced longevity and fecundity compared with controls. Remarkably, a chitinase-encoding gene was found to be laterally transferred from Microsporidia into the genomes of three different chalcid parasitoids (*N. vitripennis*, *Trichogramma prestiosum* and *Copidosoma floridanum*) (Martinson *et al.*, 2016).

Microsporidia are known to use vertical or horizontal transmission routes. There is a strong association between high pathogenicity and horizontally transmitted species (Dunn and Smith, 2001). Since microsporidia-associated transcripts were only recovered from the wild adult female transcriptome library, it is possible that *M. spermotrophus* is susceptible to a horizontally transmitted microsporidium acquired from the environment. Alternatively, microsporidian infection may be associated with the sample collection location, since the wild adult females were all collected from a natural stand of Douglas-fir, while the larvae, adult males and laboratory-reared females were all collected from infested seed harvested from a seed orchard located 20 km away. In order to determine whether microsporidia are vertically transmitted, larvae from the wild females could be screened for the presence of the parasite using targeted PCR.

There is potential to develop this microsporidium as a biological control agent against *M. spermotrophus*, since other microsporidia have been used as biological control agents in the past. Notably, microsporidia have been introduced to control such Lepidoptera as the gypsy moth (Weiser and Novotný, 1987) and the spruce budworm (Wilson and Kaupp, 1976). Microsporidia have been investigated as potential biological control agents against invasive red fire ants, *Solenopsis invicta* (Oi and Williams, 2003; Oi *et al.*, 2005). Further study is needed to establish whether microsporidia can be used as an effective control against the spread of *Megastigmus*.

7.4.3 Identification of a novel positive single-stranded RNA virus

The whole-body transcriptome of *M. spermotrophus* revealed a novel 12 kb (+)ssRNA virus. Invertebrate transcriptomes have been found to contain a diversity of novel small RNA viruses (Shi *et al.*, 2016; de Miranda *et al.*, 2017). Mining of cDNA libraries also revealed three novel ssRNA viruses in the model parasitoid *N. vitripennis* (Oliveira *et al.*, 2010). Recently, the Abisko virus was discovered from the transcriptome of the autumnal moth, *Epirrita autumnata* (de Miranda *et al.*, 2017). The RdRp of Abisko virus shares over 30% sequence homology to the RNA virus identified in the *M. spermotrophus* transcriptome. Phylogenetic reconstruction based on the amino acid sequence of ORF-1 placed Abisko virus within the insect-infected group of Negeviruses, which have a close association with plant viruses (de Miranda *et al.*, 2017). Our phylogenetic reconstruction (based on RdRp sequence homology) clustered the Abisko and the *M. spermotrophus*-associated viruses within the Hepe-Virga clade. This includes insect-infecting RNA viruses, such as the *Drosophila*-associated Boutonnet and Bofa viruses (Shi *et al.*, 2016).

The genomic arrangement of the *M. spermotrophus*-associated RNA virus is similar to other invertebrate-infecting viruses within the Hepe-Virga clade that typically encode both viral and FtsJ-like methyltransferases, viral helicases and RNA-dependent RNA polymerases. We find no evidence for OTU-like cystein proteases or for glycoprotein-containing OFRs, which are encoded by other closely related viruses (Shi *et al.*, 2016). A second transcript variant with high sequence homology to the 12 kb (+)ssRNA virus genome differed only by several single nucleotide polymorphisms (SNPs) that were mostly located within the capsid protein and RdRp domains. This transcript variant may even represent the outcome of expression from a duplicated segment of viral genome transposed into the genome of *M. spermotrophus* through horizontal gene transfer. Braconid and ichneumonid wasps encode 'domesticated' polydnaviruses within their genomes that provide the wasp with an effective delivery mechanism for viral genes essential for the wasp's offspring to develop successfully on or within the host (Burke and Strand, 2012). Integration of (+)ssRNA viral genes into the *M. spermotrophus* genome could be investigated using arbitrary PCR or genomic sequencing.

Although RNA viruses have been identified from insect transcriptome data, there are few reports of specific viral expression patterns within the host, and those few reports show differing patterns. For example, the Abisko virus was only found from a few adults and a single larval transcriptome, which led de Miranda *et al.* (2017) to conclude that the virus was unlikely to affect overall host moth population dynamics. Since our study used pooled RNA samples from multiple individual insects, the infection rate of the (+)ssRNA virus is unknown in *M. spermotrophus*. However, we did find that (+)ssRNA virus transcripts were more than four times as likely to be highly expressed in female *M. spermotrophus* compared with males and larvae. This significantly higher expression of the (+)ssRNA virus in female wasps suggests that viral particles may be co-injected with the eggs during oviposition. This was found to be the case with a different parasitic wasp, *Diachasmimorpha longicaudata* (Braconidae). *D. longicaudata* carries a symbiotic negative-sense ssRNA virus called DlRhV that is transmitted beneath the chorion of the egg during oviposition (Lawrence and Matos, 2005). DlRhV encodes a 'parasitism-specific protein' known as PSP24 that is thought to contribute to the successful parasitism of tephritid fruit flies (Simmonds *et al.*, 2016).

Another fascinating example of symbiosis between RNA viruses and parasitic wasps involves *Diadromus pulchellus* (Ichneumonidae), an insect that harbors a commensal double-stranded RNA reovirus (DpRV-1) as well as a symbiotic ascovirus (DpAV-4), which is essential for suppression of the immune system of the moth *Acrolepiosis assectella* (Bigot *et al.*, 1997; Renault *et al.*, 2002). While DpRV-1 was not found to affect the parasitic success of *D. pulchellus* directly, it has been suggested that DpRV-1 may delay the development of DpAV-4 within the wasp larvae (Renault *et al.*, 2005). The authors speculated that the capsid protein of DpRV-1 may interact with DpAV-4, as capsid proteins of other non-insect reoviruses have been shown to interfere with their host's immune system (Labrada *et al.*, 2002; Forrest and Dermody, 2003). Interestingly, *M. spermotrophus* females highly express two variants of the (+)ssRNA viral capsid protein, one encoded within a 12 kb viral genome and another found on a 2.5 kb transcript variant that contains several SNPs. These findings provide an opportunity for promising future research into potential symbiosis between a (+)ssRNA virus and a seed-feeding chalcid wasp.

7.5 Conclusions

The use of whole-body transcriptomics provided a wealth of novel discoveries in *M. spermotrophus*. This work provides evidence that a seed-feeding parasitic wasp expresses putative venom proteins. The transcriptome was also able to identify *Ralstonia* bacteria and two novel associates: a potentially pathogenic microsporidium and a potentially symbiotic (+)ssRNA virus. These may play various roles in plant host manipulation. Such findings help to further our understanding of a cryptic seed-feeding chalcid and will help us to generate new ideas and directions for future studies into host manipulation in this fascinating system. This work also provides a solid example of how transcriptomics can be a powerful high-throughput and unbiased tool for biological exploration of cryptic invasive species such as *M. spermotrophus*, but the methods described here could surely be applied as a universal discovery tool in the study of other cryptic or emerging pests, driving more targeted hypothesis testing or rapid development of biological control agents in hard-to-study systems.

Acknowledgements

This work was jointly funded by the Strategic Project Grants Partnership Program of NSERC and the Agence Nationale de Recherche – Programme Blanc International. Additional support was provided from the Canadian Institute for Advanced Research, Integrated Microbial Biodiversity Program. Researchers from the Institut National de la Recherche Agronomique (France) and Natural Resources Canada contributed to this work by sharing their expertise and providing mentorship. These included Drs Thomas Boivin, Marie-Anne Auger-Rozenberg, Alain Roques and Jean-Noël Candau. Support provided by WestGrid (www.westgrid.ca) and Compute Canada Calcul Canada (www.computecanada.ca), including that of Dr Belaid Moa from the University of Victoria were important in the *de novo* transcriptome assembly. Microsporidia-specific primer sequence was generously provided by Drs Naomi Fast and Alex Ardila-Garcia of the University of British Columbia. Infested seeds were generously provided by Dave Kolotelo of the BC Ministry of Forests, Lands and Natural Resource Operations' Tree Seed Centre, in Surrey, British Columbia. Our appreciation also goes to Madeline Antony, who corrected the references.

References

Auger-Rozenberg, M.-A., Kerdelhué, C., Magnoux, E., Turgeon, J., Rasplus, J.-Y. *et al.* (2006) Molecular phylogeny and evolution of host-plant use in conifer seed chalcids in the genus *Megastigmus* (Hymenoptera: Torymidae). *Systematic Entomology* 31(1), 47–64. DOI: 10.1111/j.1365-3113.2005.00310.x.

Auger-Rozenberg, M.-A., Boivin, T., Magnoux, E., Courtin, C., Roques, A. *et al.* (2012) Inferences on population history of a seed chalcid wasp: invasion success despite a severe founder effect from an unexpected source population. *Molecular Ecology* 21(24), 6086–6103. DOI: 10.1111/mec.12077.

Auger-Rozenberg, M.-A. and Roques, A. (2012) Seed wasp invasions promoted by unregulated seed trade affect vegetal and animal biodiversity. *Integrative Zoology* 7(3), 228–246. DOI: 10.1111/j.1749-4877.2012.00308.x.

Austin, A.D. and Dangerfield, P.C. (1998) Biology of *Mesostoa kerri* (Insecta: Hymenoptera: Braconidae: Mesostoinae), an endemic Australian wasp that causes stem galls on *Banksia marginata*. *Australian Journal of Botany* 46, 559–569.

Baker, M.D., Vossbrinck, C.R., Didier, E.S., Maddox, J.V. and Shadduck, J.A. (1995) Small subunit ribosomal DNA phylogeny of various microsporidia with emphasis on AIDS related forms. *The Journal of Eukaryotic Microbiology* 42(5), 564–570. DOI: 10.1111/j.1550-7408.1995.tb05906.x.

Becnel, J.J. and Geden, C.J. (1994) Description of a new species of microsporidia from *Muscidifurax raptor* (Hymenoptera: Pteromalidae), a pupal parasitoid of muscoid flies. *The Journal of Eukaryotic Microbiology* 41(3), 236–243. DOI: 10.1111/j.1550-7408.1994.tb01504.x.

Bigot, Y., Rabouille, A., Doury, G., Sizaret, P.Y., Delbost, F. *et al.* (1997) Biological and molecular features of the relationships between *Diadromus pulchellus* ascovirus, a parasitoid hymenopteran wasp (*Diadromus pulchellus*) and its lepidopteran host, *Acrolepiopsis assectella*. *Journal of General Virology* 78(5), 1149–1163. DOI: 10.1099/0022-1317-78-5-1149.

Boivin, T. and Candau, J.-N. (2007) Flow cytometric analysis of ploidy levels in two seed-infesting *Megastigmus* species: applications to sex ratio and species determination at the larval stage. *Entomologia Experimentalis et Applicata* 124(1), 125–131. DOI: 10.1111/j.1570-7458.2007.00576.x.

Boivin, T., Candau, J., Roques, A. and Auger-Rozenberg, M.-A. (2008) First finding of *Wolbachia* in thelytokous seed chalcids (Hymenoptera: Torymidae, *Megastigmus* sp). *Paper presented at: XXIII International Congress of Entomology 2008*, Durban, South Africa.

Boivin, T., Henri, H., Vavre, F., Gidoin, C., Veber, P. *et al.* (2014) Epidemiology of asexuality induced by the endosymbiotic *Wolbachia* across phytophagous wasp species: host plant specialization matters. *Molecular Ecology* 23(9), 2362–2375. DOI: 10.1111/mec.12737.

Bolger, A.M., Lohse, M. and Usadel, B. (2014) Trimmomatic: a flexible trimmer for Illumina sequence data. *Bioinformatics* 30(15), 2114–2120. DOI: 10.1093/bioinformatics/btu170.

Burke, G.R. and Strand, M.R. (2012) Polydnaviruses of parasitic wasps: domestication of viruses to act as gene delivery vectors. *Insects* 3(1), 91–119. DOI: 10.3390/insects3010091.

Camacho, C., Coulouris, G., Avagyan, V., Ma, N., Papadopoulos, J. *et al.* (2009) BLAST+: architecture and applications. *BMC Bioinformatics* 10(1), 1. DOI: 10.1186/1471-2105-10-421.

Capella-Gutiérrez, S., Marcet-Houben, M. and Gabaldón, T. (2012) Phylogenomics supports microsporidia as the earliest diverging clade of sequenced fungi. *BMC Biology* 10(1), 47. DOI: 10.1186/1741-7007-10-47.

Casteels, P., Ampe, C., Jacobs, F. and Tempst, P. (1993) Functional and chemical characterization of Hymenoptaecin, an antibacterial polypeptide that is infection-inducible in the honeybee (*Apis mellifera*). *Journal of Biological Chemistry* 268, 7044–7054.

Chiwocha, S., Rouault, G., Abrams, S. and von Aderkas, P. (2007) Parasitism of seed of Douglas fir (*Pseudotsuga menziesii*) by the seed chalcid, *Megastigmus spermotrophus*, and its influence on seed hormone physiology. *Sexual Plant Reproduction* 20(1), 19–25. DOI: 10.1007/s00497-006-0039-z.

de Graaf, D.C., Aerts, M., Brunain, M., Desjardins, C.A., Jacobs, F.J. *et al.* (2010) Insights into the venom composition of the ectoparasitoid wasp *Nasonia vitripennisfrom* bioinformatic and proteomic studies. *Insect Molecular Biology* 19, 11–26.

de Miranda, J., Hedman, H., Onorati, P., Stephan, J., Karlberg, O. *et al.* (2017) Characterization of a novel RNA virus discovered in the Autumnal moth *Epirrita autumnata* in Sweden. *Viruses* 9(8), 1–2.

Drezen, J.-M., Leobold, M., Bézier, A., Huguet, E., Volkoff, A.-N. *et al.* (2017) Endogenous viruses of parasitic wasps: variations on a common theme. *Current Opinion in Virology* 25, 41–48. DOI: 10.1016/j.coviro.2017.07.002.

Dunn, A.M. and Smith, J.E. (2001) Microsporidian life cycles and diversity: the relationship between virulence and transmission. *Microbes and Infection* 3(5), 381–388. DOI: 10.1016/S1286-4579(01)01394-6.

Eggleton, P. and Belshaw, R. (1992) Insect parasitoids: an evolutionary overview. *Philosophical Transactions of the Royal Society Series B Biological Sciences* 337, 1–20.

Fabre, J.-P., Auger-Rozenberg, M.-A., Chalon, A., Boivin, S. and Roques, A. (2004) Competition between exotic and native insects for seed resources in trees of a Mediterranean forest ecosystem. *Biological Invasions* 6(1), 11–22. DOI: 10.1023/B:BINV.0000010119.51131.19.

Folmer, O., Black, M., Hoeh, W., Lutz, R. and Vrijenhoek, R. (1994) DNA primers for amplification of mitochondrial cytochrome c oxidase subunit I from diverse metazoan invertebrates. *Molecular Marine Biology and Biotechnology* 3, 294–299.

Forbes, A. and Lehmann, R. (1998) Nanos and Pumilio have critical roles in the development and function of *Drosophila* germline stem cells. *Development* 125, 679–690.

Forrest, J.C. and Dermody, T.S. (2003) Reovirus receptors and pathogenesis. *Journal of Virology* 77(17), 9109–9115. DOI: 10.1128/JVI.77.17.9109-9115.2003.

Fries, I. (1993) *Nosema apis* —a parasite in the honey bee colony. *Bee World* 74(1), 5–19. DOI: 10.1080/0005772X.1993.11099149.

Fries, I., Feng, F., da Silva, A., Slemenda, S.B. and Pieniazek, N.J. (1996) *Nosema ceranae* n. sp. (Microspora, Nosematidae), morphological and molecular characterization of a microsporidian parasite of the Asian honey bee *Apis cerana* (Hymenoptera, Apidae). *European Journal of Protistology* 32(3), 356–365. DOI: 10.1016/S0932-4739(96)80059-9.

Geden, C.J., Ferreira De Almeida, M.A., Pires, A. and Prado, D. (2003) Effects of *Nosema* disease on fitness of the parasitoid *Tachinaephagus zealandicus* (Hymenoptera: Encyrtidae). *Environonmental Entomology* 32, 1139–1145.

Goto, A., Matsushita, K., Gesellchen, V., Chamy, L.E., Kuttenkeuler, D. *et al.* (2008) Akirins are highly conserved nuclear proteins required for NF-κB-dependent gene expression in *Drosophila* and mice. *Nature Immunology* 9(1), 97–104. DOI: 10.1038/ni1543.

Grissell, E.E. (1999) An annotated catalog of World Megastigminae (Hymenoptera: Chalcidoidea: Torymidae). *Contributions of the American Entomological Institute* 31, 1–92.

Guidugli, K.R., Piulachs, M.-D., Bellés, X., Lourenço, A.P. and Simões, Z.L.P. (2005) Vitellogenin expression in queen ovaries and in larvae of both sexes of *Apis mellifera*. *Archives of Insect Biochemistry and Physiology* 59(4), 211–218. DOI: 10.1002/arch.20061.

Hansen, K.D., Irizarry, R.A. and Wu, Z. (2012) Removing technical variability in RNA-Seq data using conditional quantile normalization. *Biostatistics* 13(2), 204–216. DOI: 10.1093/biostatistics/kxr054.

Herniou, E.A., Huguet, E., Thézé, J., Bézier, A., Periquet, G. *et al.* (2013) When parasitic wasps hijacked viruses: genomic and functional evolution of polydnaviruses. *Philosophical Transactions of the Royal Society B: Biological Sciences* 368(1626), 20130051. DOI: 10.1098/rstb.2013.0051.

Hirt, R.P., Logsdon, J.M., Healy, B., Dorey, M.W., Doolittle, W.F. *et al.* (1999) Microsporidia are related to fungi: evidence from the largest subunit of RNA polymerase II and other proteins. *Proceedings of the National Academy of Sciences* 96(2), 580–585. DOI: 10.1073/pnas.96.2.580.

Hoffmann, J.A. and Hetru, C. (1992) Insect defensins: inducible antibacterial peptides. *Immunology Today* 13(10), 411–415. DOI: 10.1016/0167-5699(92)90092-L.

Kelley, L.A., Mezulis, S., Yates, C.M., Wass, M.N. and Sternberg, M.J.E. (2015) The Phyre2 web portal for protein modeling, prediction and analysis. *Nature Protocols* 10(6), 845–858. DOI: 10.1038/nprot.2015.053.

LaSalle, J. and Gauld, I. (1991) Parasitic Hymenoptera and the biodiversity crisis. *Redia* 74, 515–334.

Kim, Y.S., Ryu, J.H., Han, S.J., Choi, K.H., Nam, K.B. *et al.* (2000) Gram-Negative bacteria-binding protein, a pattern recognition receptor for lipopolysaccharide and beta-1,3-glucan that mediates the signaling for the induction of innate immune genes in *Drosophila melanogaster* cells. *Journal of Biological Chemistry* 275(42), 32721–32727. DOI: 10.1074/jbc.M003934200.

Labrada, L., Bodelón, G., Viñuela, J. and Benavente, J. (2002) Avian reoviruses cause apoptosis in cultured cells: viral uncoating, but not viral gene expression, is required for apoptosis induction. *Journal of Virology* 76(16), 7932–7941. DOI: 10.1128/JVI.76.16.7932-7941.2002.

Lawrence, P.O. and Matos, L.F. (2005) Transmission of the *Diachasmimorpha longicaudata* rhabdovirus (DlRhV) to wasp offspring: an ultrastructural analysis. *Journal of Insect Physiology* 51(2), 235–241. DOI: 10.1016/j.jinsphys.2005.01.002.

Lehmann, R. and Nüsslein-Volhard, C. (1986) Abdominal segmentation, pole cell formation, and embryonic polarity require the localized activity of oskar, a maternal gene in *Drosophila*. *Cell* 47(1), 141–152. DOI: 10.1016/0092-8674(86)90375-2.

Li, B. and Dewey, C.N. (2011) RSEM: accurate transcript quantification from RNA-Seq data with or without a reference genome. *BMC Bioinformatics* 12(1), 323. DOI: 10.1186/1471-2105-12-323.

Li, W. and Godzik, A. (2006) Cd-hit: a fast program for clustering and comparing large sets of protein or nucleotide sequences. *Bioinformatics* 22(13), 1658–1659. DOI: 10.1093/bioinformatics/btl158.

Martinson, E.O., Martinson, V.G., Edwards, R., Mrinalini, M. and Werren, J.H. (2016) Laterally transferred gene recruited as a venom in parasitoid wasps. *Molecular Biology and Evolution* 33(4), 1042–1052. DOI: 10.1093/molbev/msv348.

Moran, N.A. and Plague, G.R. (2004) Genomic changes following host restriction in bacteria. *Current Opinion Genetic Development* 14(6), 627–633. DOI: 10.1016/j.gde.2004.09.003.

Munro, J.B., Heraty, J.M., Burks, R.A., Hawks, D., Mottern, J. *et al.* (2011) A molecular phylogeny of the Chalcidoidea (Hymenoptera). *PLoS ONE* 6(11), e27023. DOI: 10.1371/journal.pone.0027023.

Newton, I.L.G. and Bordenstein, S.R. (2011) Correlations between bacterial ecology and mobile DNA. *Current Microbiology* 62(1), 198–208. DOI: 10.1007/s00284-010-9693-3.

Oi, D.H. and Williams, D.F. (2003) (Microsporidia: Thelohaniidae) Infection in reproductives of red imported fire ants (Hymenoptera: Formicidae) and its implication for intercolony transmission. *Environmental Entomology* 32(5), 1171–1176. DOI: 10.1603/0046-225X-32.5.1171.

Oi, D.H., Briano, J.A., Valles, S.M. and Williams, D.F. (2005) Transmission of *Vairimorpha invictae* (Microsporidia: Burenellidae) infections between red imported fire ant (Hymenoptera: Formicidae) colonies. *Journal of Invertebrate Pathology* 88(2), 108–115. DOI: 10.1016/j.jip.2004.11.006.

Oliveira, D.C.S.G., Hunter, W.B., Ng, J., Desjardins, C.A., Dang, P.M. *et al.* (2010) Data mining cDNAs reveals three new single stranded RNA viruses in *Nasonia* (Hymenoptera: Pteromalidae). *Insect Molecular Biology* 19, 99–107. DOI: 10.1111/j.1365-2583.2009.00934.x.

Paulson, A.R., von Aderkas, P. and Perlman, S.J. (2014) Bacterial associates of seed-parasitic wasps (Hymenoptera: *Megastigmus*). *BMC Microbiology* 14(1), 224. DOI: 10.1186/s12866-014-0224-4.

Paulson, A.R., Le, C.H., Dickson, J.C., Ehlting, J., von Aderkas, P. *et al.* (2016) Transcriptome analysis provides insight into venom evolution in a seed-parasitic wasp, *Megastigmus spermotrophus*. *Insect Molecular Biology* 25(5), 604–616. DOI: 10.1111/imb.12247.

Pelosi, P., Zhou, J.-J., Ban, L.P. and Calvello, M. (2006) Soluble proteins in insect chemical communication. *Cellular and Molecular Life Sciences* 63(14), 1658–1676. DOI: 10.1007/s00018-005-5607-0.

Pertea, G., Huang, X., Liang, F., Antonescu, V., Sultana, R. *et al.* (2003) TIGR gene indices clustering tools (TGICL): a software system for fast clustering of large EST datasets. *Bioinformatics* 19(5), 651–652. DOI: 10.1093/bioinformatics/btg034.

RStudio Team (2014) RStudio: Integrated development for R. RStudio, Inc., Boston, Massachusetts. Available at: http: //www.rstudio.com/

Renault, S., Petit, A., Bénédet, F., Bigot, S. and Bigot, Y. (2002) Effects of the *Diadromus pulchellus* ascovirus, DpAV-4, on the hemocytic encapsulation response and capsule melanization of the leek-moth pupa, *Acrolepiopsis assectella*. *Journal of Insect Physiology* 48(3), 297–302. DOI: 10.1016/S0022-1910(01)00174-3.

Renault, S., Stasiak, K., Federici, B. and Bigot, Y. (2005) Commensal and mutualistic relationships of reoviruses with their parasitoid wasp hosts. *Journal of Insect Physiology* 51(2), 137–148. DOI: 10.1016/j.jinsphys.2004.08.002.

Robertson, G., Schein, J., Chiu, R., Corbett, R., Field, M. *et al.* (2010) *De novo* assembly and analysis of RNA-seq data. *Nature Methods* 7(11), 909–912. DOI: 10.1038/nmeth.1517.

Ronquist, F. and Liljeblad, J. (2001) Evolution of the gall wasp-host plant association. *Evolution* 55(12), 2503–2522. DOI: 10.1111/j.0014-3820.2001.tb00765.x.

Roques, A. and Skrzypczyńska, M. (2003) Seed-infesting chalcids of the genus *Megastigmus* Dalman, 1820 (Hymenoptera: Torymidae) native and introduced to the West Palearctic region: taxonomy, host specificity and distribution. *Journal of Natural History* 37(2), 127–238. DOI: 10.1080/713834669.

Roques, A., Sun, J.-H., Auger-Rozenberg, M.-A. and Hua, O. (2003) Potential invasion of China by exotic insect pests associated with tree seeds. *Biodiversity and Conservation* 12(11), 2195–2210. DOI: 10.1023/A:1024592716172.

Schonbaum, C.P., Perrino, J.J. and Mahowald, A.P. (2000) Regulation of the vitellogenin receptor during *Drosophila melanogaster* oogenesis. *Molecular Biology of the Cell* 11(2), 511–521. DOI: 10.1091/mbc.11.2.511.

Shi, M., Lin, X.-D., Tian, J.-H., Chen, L.-J., Chen, X. *et al.* (2016) Redefining the invertebrate RNA virosphere. *Nature* 540(7634), 539–543. DOI: 10.1038/nature20167.

Simmonds, T.J., Carrillo, D. and Burke, G.R. (2016) Characterization of a venom gland-associated rhabdovirus in the parasitoid wasp *Diachasmimorpha longicaudata*. *Journal of Insect Physiology* 91–92, 48–55. DOI: 10.1016/j.jinsphys.2016.06.009.

Solter, L.F., Becnel, J.J. and Oi, D.H. (2012) Microsporidian entomophathogens. In: Vega, F.E. and Kaya, H.K. (eds) *Insect Pathology*. Academic Press, Waltham, Massachusetts, pp. 221–263.

Tamura, K., Peterson, D., Peterson, N., Stecher, G., Nei, M. *et al.* (2011) MEGA5: molecular evolutionary genetics analysis using maximum likelihood, evolutionary distance, and maximum parsimony methods. *Molecular Biology and Evolution* 28(10), 2731–2739. DOI: 10.1093/molbev/msr121.

Tarazona, S., García-Alcalde, F., Dopazo, J., Ferrer, A. and Conesa, A. (2011) Differential expression in RNA-Seq: a matter of depth. *Genome Research* 21(12), 2213–2223. DOI: 10.1101/gr.124321.111.

Turgeon, J.J., Roques, A. and De Groot, P. (1994) Insect fauna of coniferous seed cones: diversity, host plant interactions, and management. *Annual Review of Entomology* 39(1), 179–212. DOI: 10.1146/annurev.en.39.010194.001143.

Vogt, R.G., Callahan, F.E., Rogers, M.E. and Dickens, J.C. (1999) Odorant binding protein diversity and distribution among the insect orders, as indicated by LAP, an OBP-related protein of the true bug *Lygus lineolaris* (Hemiptera, Heteroptera). *Chemical Senses* 24(5), 481–495. DOI: 10.1093/chemse/24.5.481.

von Aderkas, P., Rouault, G., Wagner, R., Chiwocha, S. and Roques, A. (2005a) Multinucleate storage cells in Douglas-fir (*Pseudotsuga menziesii* (Mirbel) Franco) and the effect of seed parasitism by the chalcid *Megastigmus spermotrophus* Wachtl. *Heredity* 94(6), 616–622. DOI: 10.1038/sj.hdy.6800670.

von Aderkas, P., Rouault, G., Wagner, R., Rohr, R. and Roques, A. (2005b) Seed parasitism redirects ovule development in Douglas fir. *Proceedings of the Royal Society B: Biological Sciences* 272(1571), 1491–1496. DOI: 10.1098/rspb.2005.3061.

Vossbrinck, C.R. and Debrunner-Vossbrinck, B.A. (2005) Molecular phylogeny of the microsporidia: ecological, ultrastructural and taxonomic considerations. *Folia Parasitologica* 52(1), 131–142. DOI: 10.14411/fp.2005.017.

Weiser, J. and Novotný, J. (1987) Field application of *Nosema lymantriae* against the gypsy moth, *Lymantria dispar* L. *Journal of Applied Entomology* 104(1-5), 58–62. DOI: 10.1111/j.1439-0418.1987.tb00496.x.

Werren, J.H., Richards, S., Desjardins, C.A., Niehuis, O., Gadau, J. *et al.* (2010) Functional and evolutionary insights from the genomes of three parasitoid *Nasonia* species. *Science* 327(5963), 343–348. DOI: 10.1126/science.1178028.

Wilson, G. and Kaupp, J. (1976) Application of *Nosema fumiferanae* and *Pleistophora schubergi* (Microsporidia) against the spruce budworm in Ontario, 1976. *Inf. Rep* IP-X-15(1–14) Canadian Forest Service Sault Ste. Marie, Ontario.

Yang, Y. and Smith, S.A. (2013) Optimizing de novo assembly of short-read RNA-Seq data for phylogenomics. *BMC Genomics* 14(1), 1. DOI: 10.1186/1471-2164-14-328.

Zchori-Fein, E., Geden, C.J. and Rutz, D.A. (1992) Microsporidioses of *Muscidifurax raptor* (Hymenoptera: Pteromalidae) and other pteromalid parasitoids of muscoid flies. *Journal of Invertebrate Pathology* 60(3), 292–298. DOI: 10.1016/0022-2011(92)90011-R.

8 Differential Transcriptome Profiling for Identification of Cellulose Degrading Enzymes in *Ctenolepisma longicaudata*

Ratnasri Pothula,[1*] Brian R. Johnson,[2] William E. Klingeman,[3] Matthew Huff,[1] Margaret E. Staton,[1] and Juan Luis Jurat-Fuentes[1]

[1]*Department of Entomology and Plant Pathology, University of Tennessee, Knoxville, Tennessee, USA;*
[2]*Department of Entomology and Nematology, University of California, Davis, California, USA;*
[3]*Department of Plant Sciences, University of Tennessee, Knoxville, Tennessee, USA*

8.1 Introduction

Whole-transcriptome shotgun sequencing, or RNA-seq, is a next-generation sequencing technique that enables quantitative determination of total transcripts present to serve as a proxy for the level of expression of the corresponding gene in a cell, tissue, or whole organism (Wang *et al.*, 2009). Compared with microarrays and other hybridization-based approaches, RNA-seq does not suffer from cross-hybridization bias and is not limited by availability of genomic resources for the organism studied (Ekblom and Galindo, 2011), making this technique very attractive in studying non-model organisms for which genomic data are scarce. Additionally, comparative RNA-seq studies allow for gene expression profiling between different tissues and/or conditions, facilitating the identification of genes with specific functional attributes.

Interest in the identification of plant cell wall degrading enzyme (PCWDE) genes has increased in recent years with the need to identify novel cellulase enzymes to optimize plant biomass conversion to glucose (saccharification)

*Corresponding author: rmallipe@vols.utk.edu

in ethanol biofuel production (Gomez *et al.*, 2008). Cellulases are a class of PCWDEs that completely digest plant cellulose into glucose and include endoglucanases that cleave the cellulose chain internally at random locations, exoglucanases that cleave the cellulose chain at the ends to release cellobiose, and β-glucosidases that degrade cellobiose into glucose units (Watanabe and Tokuda, 2010). Cellulases have been traditionally described from microorganisms, yet in the past decade insects have also been proposed as a prospecting resource for PCWDEs with unique properties, including cellulases (endoglucanases and β-glucosidases), hemicellulases, and pectinases (Watanabe and Tokuda, 2010; Calderón-Cortés *et al.*, 2012). The lack to date of described insect exoglucanases may be explained by a higher number of endoglucanase insect enzymes with dual endo/exo activity (Scrivener and Slaytor, 1994), physical processing by mandibles and proventriculus, and/or a long digestive tract that allows for increased length of digestion in the gut (Watanabe and Tokuda, 2010; Calderón-Cortés *et al.*, 2012). In contrast to cellulase specificity for cellulose, hemicellulases and pectinases digest hemicellulose and pectin polysaccharides, respectively, which are interlocked with cellulose fibers in the plant cell wall (Gilbert, 2010). Insects belonging to 16 taxonomic orders were reported to have endogenous production of one or more of these PCWDEs (Calderón-Cortés *et al.*, 2012), yet molecular evidence has only been provided for insects in eight of these orders. From these studies, molecular evidence has been provided for endogenous PCWDEs, such as endoglucanases belonging to glycoside hydrolase (GH) families 9 and 45, β-glucosidases of GH family 5, hemicellulases of GH families 5 and 11, β–1,3-glucanases of GH family 16, and pectinases of GH family 28 (Calderón-Cortés *et al.*, 2012).

Zygentoma is a primitive non-model insect group recently characterized as displaying the highest relative cellulase activity compared with other insect groups, including model insects traditionally used in cellulose degradation research, such as termites, cockroaches, and beetles (Pothula *et al.*, 2019). Previous work detected endoglucanase, β-glucosidase, xylanase, pectinase, amylase, maltase, sucrase, and lactase activities in foregut fluids from firebrats (*Thermobia domestica*) and silverfish (*Ctenolepisma longicaudata*) through biochemical tests (Zinkler and Götze, 1987; Pothula *et al.*, 2019). In addition, investigation of the digestive proteome of *T. domestica* revealed the production of lytic polysaccharide mono-oxygenases (LPMOs), which are enzymes that form abrasions or small cuts in cellulose fibers, thereby making the fibers more accessible for breakdown by cellulases (Sabbadin *et al.*, 2018). However, there is a dearth of gene expression information as it relates to cellulolytic activity in these non-model insects.

We have used comparative RNA-seq to identify cellulase genes and their differential expression in the foregut of the silverfish *C. longicaudata*, as the foregut is the main location of cellulolytic activity within the digestive tract of this insect (Pothula *et al.*, 2019). Transcriptomes from insects fed on four diets consisting of different cellulosic substrates were compared to characterize the relationship between PCWDE gene expression and diet. Differential gene expression analysis revealed that cellulase gene expression is primarily driven by type of tissue rather than diet, yet when compared with rest-of-body tissues, foregut tissues presented a higher number of PCWDEs that were significantly up-regulated when *C. longicaudata* were fed on the diet containing the highest cellulose content (paper). Here, we explain how the annotation of differentially expressed PCWDE genes has revealed highest identity to insect homologs, which in turn supports the potential conservation of insect PCWDEs through evolution. Overall, our research demonstrates the utility of RNA-seq to identify candidate functional PCWDE and LPMO genes from primitive insects.

8.2 Materials and Methods

8.2.1 Insect rearing

Several batches of nymphs and adults of silverfish *C. longicaudata* were trapped into lidless plastic dishes baited with whole-grain oat flakes (Quick 1 minute Oats, Quaker Oats Co., Chicago, Illinois) in buildings at the campus of the University of Tennessee (Knoxville, Tennesee). These insects were then reared in the laboratory at room temperature on rolled oats, paper, and Pedigree® Adult Complete Nutrition dry dog food (Mars Inc., Mount Olive, New Jersey).

8.2.2 Diet treatments

Six adult *C. longicaudata* per treatment were starved for 5 days in individual Petri dishes. On the sixth day, six randomly selected insects were provided with one of four diet treatments and allowed to feed for 5 days. Treatments included 0.2% carboxymethylcellulose (CMC) dissolved in 1% agar, standard 92 multipurpose printing paper (Georgia-Pacific, Atlanta, Georgia), pulverized switchgrass (generously provided by Dr Nicole Labbe, Center for Renewable Carbon, University of Tennessee), and a protein-balanced diet (NatureWise® Chick Starter Grower Feed, Nutrena-Cargill Mills, Giddings, Texas) as a control. A block of 1% agar was provided as a water source to all insects. Diet and agar blocks were changed every other day. Each treatment was replicated three times.

8.2.3 Dissections, RNA extraction and sequencing

Three replicates of six fed insects per treatment were anesthetized by incubation at 4°C for 10 min. The foregut was quickly dissected from the rest of the body by pulling from the head with forceps and then cutting the head from the foregut with a disposable scalpel. Individual foreguts and rest-of-body (carcass including head) were collected into separate microcentrifuge tubes, snap frozen in liquid nitrogen and stored at –80°C. Total RNA was extracted from pools of six foreguts or rest-of-body samples using Direct-zol RNA MiniPrep Plus (Zymo Research, Irvine, California) and RNeasy Maxi kits (Qiagen, Hilden, Germany), respectively. The use of different kits depending on tissue was necessary to accommodate differences in tissue weights. Purity of total RNA was checked using NanoDrop and then submitted to the Genomic Sequencing and Analysis Facility at the University of Texas at Austin (Texas) for library preparation and sequencing. Total RNA quality was analyzed using a Bioanalyzer (Agilent Technologies, Santa Clara, California), and then enriched for poly-A mRNA using the Poly(A) Purist MAG kit (Thermo Fisher Scientific, Waltham, Massachusetts), according to manufacturer's instructions. Libraries were prepared with the NEBNext Ultra II Directional RNA Library Prep kit for Illumina (Ipswich, Massachusetts) before paired-end sequencing (300 base pairs) on a Hi-Seq 4000 (Illumina, Inc., San Diego, California).

8.2.4 Differential gene expression analysis

A draft genome of *C. longicaudata* was assembled following the same bioinformatic pipeline used by Brand *et al.* (2018) for *T. domestica* (B. Johnson, unpublished). Coding sequences from the draft genome were extracted as a FASTA file using MAKER v2.31.9 (Cantarel *et al.*, 2008) and indexed using Salmon v0.8.2 (Patro *et al.*, 2017). The abundance of RNA-seq reads was quantified using Salmon due to similar accuracy but increased speed when compared with similar quantification programs. The resulting read of raw counts per transcript in each replicate was used as input to detect the differentially expressed genes between treatments using DESeq2 (Love *et al.*, 2014). The DESeq2 program is an R package that uses negative binomial generalized linear models to test for differentially expressed genes (DEG) between treatments. Among the detected DEGs, significance was established at an adjusted *P*-value (P_{adj}) <0.05. Principal component analysis (PCA) plots and heatmaps were developed to visualize the differential expression of cellulase genes and LPMOs across treatments and tissues using plotPCA' and 'pheatmap' options in RStudio version 1.1.423 (RStudio, Inc. Boston, Massachusetts).

8.2.5 Annotation and alignment analysis of PCWDEs

Differentially expressed cellulase and LPMO genes in *C. longicaudata* were identified by annotating all the DEGs (P_{adj} <0.05) between pair-wise treatment comparisons using a custom-made database in Blast2GO (https://www.blast2go.com/) that included all protein sets from NCBI with known cellulase activity from insects, bacteria, fungi, and protozoa. All the resulting DEG PCWDEs and LPMOs were manually blasted in the NCBI website against the non-redundant (nr) protein database to get the most specific match. The annotated sequences encoding each PCWDE group were aligned with the respective enzyme group protein sequences of invertebrates extracted from the Identical Protein Groups (IPG) database in NCBI using Multiple Alignment through the Fast Fourier Transform tool in Geneious Prime 2019.0.4 (http://www.geneious.com).

8.3 Results

8.3.1 Predicted PCWDEs and LPMO encoding genes in a *C. longicaudata* draft genome

Annotation of all coding sequences in a draft genome for *C. longicaudata* (B. Johnson, unpublished) resulted in identification of 69 sequences encoding for β-1,4-endoglucanases, 22 encoding for β-glucosidases, 50 encoding for α-glucosidases, 32 encoding mannanases, 30 encoding β-glucuronidases, 26 encoding β-galactosidases, 16 encoding β1,3-glucanases and 67 genes encoding for LPMOs (Table 8.1).

8.3.2 Gene expression analysis in foregut

Differential gene expression analysis in *C. longicaudata* foregut tissue fed on CMC compared with a control (protein) diet resulted in no changes in expression of cellulase genes. In contrast, feeding *C. longicaudata* on switchgrass versus control diet resulted in up-regulation of one endoglucanase (GH 9) (3.2-fold), one β-galactosidase (5.7-fold), and two β-glucuronidases in the foregut tissue (Table 8.2). However, the greatest changes were detected when comparing insects fed on paper with those fed on control diet, with significant up-regulation of 13 endoglucanases (between 1.8-fold to 4.8-fold) (Table 8.2). Similar up-regulation of endoglucanase enzymes was observed when comparing feeding on paper with CMC or switchgrass diets, resulting in 12 (2.3–4.7-fold) and five (3.10–4.42-fold) up-regulated endoglucase genes,

Table 8.1. Number of genes encoding for plant cell wall degrading enzyme (PCWDE) and lytic polysaccharide mono-oxgenases (LPMOs) in the draft *Ctenolepisma longicaudata* genome. Also shown is the number of differentially expressed sequences from our RNA-seq study.

Enzyme (glycosyl hydrolase family)	Number of genes encoding cellulases and LPMOs in the draft genome	Number of genes differentially expressed
β-1,4-endoglucanase (GH9)	69	31
β-glucosidase (GH1)	22	18
Mannanase (GH5)	32	1
α-glucosidase (GH31)	50	5
β-glucuronidase (GH2)	30	6
β-galactosidase (GH35)	26	5
β1,3-glucanase (GH16)	16	2
Lytic polysaccharide mono-oxygenases (LPMOs) (AA15)	67	34

Table 8.2. Effect of diet on plant cell wall degrading enzyme (PCWDE) gene expression in foregut tissue of *Ctenolepisma longicaudata*. Differentially expressed PCWDE genes in different pair-wise treatment comparisons of foregut tissue of *C. longicaudata* fed on carboxymethyl cellulose (CMC), paper, protein, and switchgrass diets ($P_{adj} < 0.05$).

Coding sequence	Enzyme	Up-regulated diet	Log_2-fold change	P_{adj} value
Paper vs protein				
Lep_00006775-RA	Hydrolase family 9	Paper	4.8	<0.001
Lep_00022398-RA	Hydrolase family 9	Paper	1.8	0.02
Lep_00034744-RA	Hydrolase family 9	Paper	1.9	0.04
Lep_00036184-RA	Hydrolase family 9	Paper	3.5	0.002
Lep_00044891-RA	Hydrolase family 9	Paper	3.0	0.001
Lep_00051683-RA	Hydrolase family 9	Paper	2.3	0.005
Lep_00051684-RA	Hydrolase family 9	Paper	2.3	0.002
Lep_00052326-RA	Hydrolase family 9	Paper	2.3	0.003
Lep_00060202-RA	Hydrolase family 9	Paper	2.7	0.001
Lep_00071132-RA	Hydrolase family 9	Paper	2.6	<0.001
Lep_00071754-RA	Hydrolase family 9	Paper	1.9	0.005
Lep_00115213-RA	Hydrolase family 9	Paper	2.8	0.008
Lep_00135466-RA	Hydrolase family 9	Paper	4.7	<0.001
Switchgrass vs protein				
Lep_00006757-RA	β-Galactosidase-1 3	Switchgrass	5.7	0.04
Lep_00028614-RA	β-Glucuronidase	Switchgrass	4.6	0.02
Lep_00055855-RA	β-Glucuronidase	Switchgrass	6.5	0.001
Lep_00096399-RA	Hydrolase family 9	Protein	3.2	0.02
Paper vs CMC				
Lep_00006775-RA	Hydrolase family 9	Paper	3.8	<0.001
Lep_00012530-RA	Mannanase	CMC	1.7	0.04
Lep_00022398-RA	Hydrolase family 9	Paper	2.6	0.002

Continued

Table 8.2. Continued

Coding sequence	Enzyme	Up-regulated diet	Log_2-fold change	P_{adj} value
Lep_00034832-RA	Endo-β-1,4-glucanase	Paper	2.9	0.02
Lep_00036184-RA	Hydrolase family 9	Paper	3.3	<0.001
Lep_00044891-RA	Hydrolase family 9	Paper	2.8	<0.001
Lep_00060202-RA	Hydrolase family 9	Paper	2.5	<0.001
Lep_00071132-RA	Hydrolase family 9	Paper	2.3	0.001
Lep_00071461-RA	Hydrolase family 9	Paper	4.1	<0.001
Lep_00078550-RA	Hydrolase family 9	Paper	4.7	<0.001
Lep_00104311-RA	Endo-β-1,4-glucanase	Paper	2.5	0.04
Lep_00115213-RA	Hydrolase family 9	Paper	2.9	<0.001
Lep_00135466-RA	Hydrolase family 9	Paper	3.8	<0.001
Switchgrass vs paper				
Lep_00029521-RA	Endo-β-1,4-glucanase	Paper	3.3	0.005
Lep_00036184-RA	Hydrolase family 9	Paper	3.1	0.005
Lep_00039163-RA	β1,3-Glucan-binding-like	Paper	3.8	<0.001
Lep_00071461-RA	Hydrolase family 9	Paper	3.7	<0.001
Lep_00078550-RA	Hydrolase family 9	Paper	4.4	<0.001
Lep_00135466-RA	Hydrolase family 9	Paper	3.7	<0.001

Table 8.3. Effect of diet on lytic polysaccharide monooxygenase (LPMO) gene expression in foregut tissue of *Ctenolepisma longicaudata*. Differentially expressed lytic polysaccharide monooxygenases (LPMOs) genes in different pair-wise treatment comparisons of foregut tissue of *C. longicaudata* fed on carboxymethyl cellulose (CMC), paper, protein, and switchgrass diets ($P_{adj} < 0.05$).

Coding sequence	Up-regulated diet	Log_2-fold change	P_{adj} value
Paper vs protein			
Lep_00018140-RA	Paper	3.5	0.04
Lep_00029394-RA	Paper	3.2	0.03
Lep_00037556-RA	Paper	3.5	<0.001
Lep_00039537-RA	Paper	3.9	0.006
Lep_00083016-RA	Paper	2.8	0.02
Lep_00084324-RA	Paper	2.9	0.02
Lep_00097915-RA	Paper	3.3	0.03
Lep_00103234-RA	Paper	3.1	0.02
Lep_00106678-RA	Paper	2.6	<0.001
Lep_00118447-RA	Paper	2.8	0.005
Lep_00136402-RA	Paper	3.2	0.04
Switchgrass vs protein			
Lep_00083016-RA	Switchgrass	4.2	0.005
Lep_00097788-RA	Switchgrass	5.0	0.04
Lep_00119656-RA	Switchgrass	4.4	0.04
Paper vs CMC			
Lep_00110038-RA	CMC	4.4	0.04
Switchgrass vs CMC			
Lep_00121885-RA	Switchgrass	6.8	0.02

respectively. Feeding on paper versus switchgrass also resulted in 3.8-fold up-regulation of a β-1,3-glucanase (Table 8.2).

Similar to cellulases, differential gene expression analysis revealed that in the foregut tissue of *C. longicaudata* fed on CMC compared with control diet there were no differentially expressed LPMO genes. However, 11 (2.6–3.9-fold) and three (4.2–5.0-fold) LPMO encoding genes were significantly up-regulated in the foregut tissue of *C. longicaudata* fed on paper and switchgrass, respectively, when compared with protein diet (Table 8.3). Feeding of *C. longicaudata* on CMC and switchgrass resulted in 4.4-fold and 6.8-fold up-regulation of only one LPMO gene in the foregut tissue compared with feeding on paper and CMC, respectively (Table 8.3).

8.3.3 Gene expression analysis in rest-of-body

We analyzed gene expression in the rest of the body after dissection of the foregut tissue to detect PCWDE transcripts produced in the salivary glands, midgut and hindgut. Unlike in foregut tissue, differential gene expression analysis on the rest of the body revealed that feeding *C. longicaudata* on CMC resulted in differential cellulase gene expression compared with protein diet. One glucosidase was up-regulated (1.8-fold) in protein compared with CMC diet, while two endoglucanase (1.4-fold and 2.7-fold) encoding genes, two β-galactosidases (2.7-fold and 3.5-fold) and nine glucosidases (ranging from 1.4-fold to 3.5-fold) were up-regulated in the reciprocal comparison (Table 8.4). Similarly, seven genes were up-regulated that encoded for endoglucanases (2.3–4.2-fold) and one glucosidase (3.9-fold) when *C. longicaudata* was fed on paper compared with protein diet. When comparing protein with paper diet, two endoglucanases were up-regulated (2.2-fold and 2.4-fold), as well as two glucosidases (2.0-fold and 2.3-fold), and one galactosidase (2.4-fold) (Table 8.4). Likewise, two endoglucanase encoding genes (2.9-fold and 3-fold), two β-glucosidases (2.1-fold and 3.1-fold), one β-1,3-glucanase (1.5-fold), and one gene encoding for an enzyme in GH 65 (2.4-fold) were up-regulated in *C. longicaudata* fed on switchgrass compared with protein. In the reciprocal comparison, we detected five glucosidases (1.5–5.0-fold), four endoglucanases (2.5–3.9-fold), and one β-galactosidase (2.1-fold) as up-regulated (Table 8.4). Feeding *C. longicaudata* on switchgrass up-regulated the expression of 15 glucosidases ranging from 1.3-fold to 3.5-fold, nine endoglucanases ranging from 2-fold to 3.9-fold, five β-glucuronidases (1.9–3.1-fold), one β-1,3-glucanase by 1.4-fold, one β-galactosidase by 1.8-fold, one β-1,6-glucanase by 2.6-fold, and one GH 65 enzyme by 2.9-fold in the rest-of-body sample when compared with feeding on CMC. In contrast, CMC up-regulated the expression of only three endoglucanases (3.6–3.9-fold) compared with switchgrass. As detected for foregut, the greatest changes in PCWDE gene expression were detected when *C. longicaudata* was fed paper compared with all other diets. Feeding *C. longicaudata* on a paper diet up-regulated all cellulase encoding genes, including 16 endoglucanases (1.3–4.9-fold), two β-glucosidases (3.4-fold and 3.6-fold), and one β1,6-glucanase (2.2-fold) compared with feeding on CMC (Table 8.4). Additionally, 16 endoglucanases were up-regulated (ranging from 1.4-fold to 4.4-fold) in the rest-of-body sample from *C. longicaudata* that were fed paper, when compared with those fed on switchgrass. In the reciprocal comparison, only one mannosidase (2.6-fold), one glucosidase (1.3-fold), and one β-galactosidase (1.4-fold) were up-regulated in the rest-of-body sample of *C. longicaudata* fed switchgrass in comparison with insects fed the paper diet (Table 8.4).

LPMOs were not significantly differentially expressed in the rest-of-body of *C. longicaudata* fed on CMC and paper compared with protein diet. However, six LPMO encoding genes were significantly up-regulated (2.1–2.9-fold) in paper compared with CMC diet (Table 8.5). Similarly, the expression of 18 (1.6–3.6-fold), six (1.8–3.2-fold) and five LPMOs (1.5–2.6-fold) were significantly up-regulated in the rest-of-body of insects fed on switchgrass compared with CMC, protein, and paper diets, respectively (Table 8.5). Overall, 16 genes encoding cellulases and six LPMO encoding genes were

Table 8.4. Effect of diet on plant cell wall degrading enzyme (PCWDE) gene expression in rest-of-body (carcass after dissection of foregut) sample of *Ctenolepisma longicaudata*. Differentially expressed PCWDE genes in different pair-wise treatment comparisons of rest-of-body sample of *C. longicaudata* fed on carboxymethyl cellulose (CMC), paper, protein, and switchgrass diets ($P_{adj} < 0.05$).

Coding sequence	Enzyme	Up-regulated diet	Log_2-fold change	P_{adj} value
CMC vs protein				
Lep_00026156-RA	Hydrolase family 9	CMC	1.4	0.03
Lep_00038648-RA	Hydrolase family 9	CMC	2.7	0.04
Lep_00004873-RA	Glucosidase	CMC	2.3	0.04
Lep_00011407-RA	Glucosidase	Protein	1.8	0.04
Lep_00015101-RA	Glucosidase	CMC	2.3	0.02
Lep_00020108-RA	Glucosidase	CMC	2.4	0.006
Lep_00052604-RA	Glucosidase	CMC	3.5	0.002
Lep_00097528-RA	Glucosidase	CMC	2.5	0.02
Lep_00098705-RA	Glucosidase	CMC	2.1	0.001
Lep_00079073-RA	Uncharacterized family 31 glucosidase KIAA1161-like	CMC	3.3	0.001
Lep_00094746-RA	Uncharacterized family 31 glucosidase KIAA1161-like	CMC	2.4	<0.001
Lep_00102111-RA	Uncharacterized family 31 glucosidase KIAA1161-like	CMC	3.1	<0.001
Lep_00036877-RA	β-Galactosidase-1 2	CMC	3.5	0.03
Lep_00059101-RA	β-Galactosidase-1 2	CMC	2.7	0.006
Paper vs protein				
Lep_00006775-RA	Hydrolase family 9	Paper	3.8	<0.001
Lep_00026154-RA	Endoglucanase	Protein	2.2	0.04
Lep_00026156-RA	Hydrolase family 9	Protein	2.4	<0.001
Lep_00036184-RA	Hydrolase family 9	Paper	2.4	<0.001
Lep_00042852-RA	Hydrolase family 9	Paper	2.4	<0.001

Continued

Table 8.4. Continued

Coding sequence	Enzyme	Up-regulated diet	Log_2-fold change	P_{adj} value
Lep_00060202-RA	Hydrolase family 9	Paper	2.3	0.005
Lep_00071461-RA	Hydrolase family 9	Paper	3.8	<0.001
Lep_00078550-RA	Hydrolase family 9	Paper	4.2	<0.001
Lep_00135466-RA	Hydrolase family 9	Paper	3.0	<0.001
Lep_00004872-RA	Glucosidase	Protein	2.3	0.002
Lep_00102111-RA	Uncharacterized family 31 glucosidase KIAA1161-like	Protein	2.0	0.04
Lep_00113088-RA	Glucosidase	Paper	3.9	0.008
Lep_00059101-RA	β-Galactosidase-1 2	Protein	2.4	0.04
Switchgrass vs protein				
Lep_00014981-RA	Hydrolase family 9	Protein	2.5	0.004
Lep_00014982-RA	β-1,4-Glucanase 5	Protein	2.7	0.003
Lep_00014983-RA	Hydrolase family 9	Protein	3.4	0.03
Lep_00042852-RA	Hydrolase family 9	Switchgrass	3.0	<0.001
Lep_00095117-RA	Endo-β-1,4-glucanase	Switchgrass	2.9	0.005
Lep_00096399-RA	Hydrolase family 9	Protein	3.9	<0.001
Lep_00004872-RA	Glucosidase	Protein	1.5	0.04
Lep_00021305-RA	Glucosidase	Switchgrass	2.1	0.04
Lep_00065284-RA	Glucosidase	Switchgrass	3.1	0.003
Lep_00077538-RA	Uncharacterized family 31 glucosidase KIAA1161-like	Protein	2.0	0.04
Lep_00079073-RA	Uncharacterized family 31 glucosidase KIAA1161-like	Protein	2.4	0.03
Lep_00102111-RA	Uncharacterized family 31 glucosidase KIAA1161-like	Protein	2.5	<0.001
Lep_00113088-RA	Glucosidase	Protein	5.0	<0.001

Continued

Table 8.4. Continued

Coding sequence	Enzyme	Up-regulated diet	Log_2-fold change	P_{adj} value
Lep_00036877-RA	β-Galactosidase-1 2	Protein	2.1	0.03
Lep_00034316-RA	β-1,3-Glucan-binding-like	Switchgrass	1.5	0.04
Lep_00050000-RA	Hydrolase family 65	Switchgrass	2.4	0.003
Paper vs CMC				
Lep_00006775-RA	Hydrolase family 9	Paper	4.2	<0.001
Lep_00022398-RA	Hydrolase family 9	Paper	2.1	0.001
Lep_00036184-RA	Hydrolase family 9	Paper	2.8	<0.001
Lep_00038648-RA	Hydrolase family 9	Paper	4.3	<0.001
Lep_00042851-RA	Hydrolase family 9	Paper	3.2	0.04
Lep_00044891-RA	Hydrolase family 9	Paper	2.9	<0.001
Lep_00051683-RA	Hydrolase family 9	Paper	1.9	0.004
Lep_00051684-RA	Hydrolase family 9	Paper	2.0	<0.001
Lep_00052326-RA	Hydrolase family 9	Paper	2.0	<0.001
Lep_00060202-RA	Hydrolase family 9	Paper	3.0	<0.001
Lep_00071132-RA	Hydrolase family 9	Paper	2.2	<0.001
Lep_00071461-RA	Hydrolase family 9	Paper	1.3	<0.001
Lep_00071754-RA	Hydrolase family 9	Paper	1.5	<0.001
Lep_00078550-RA	Hydrolase family 9	Paper	4.9	<0.001
Lep_00115213-RA	Hydrolase family 9	Paper	3.2	<0.001
Lep_00135466-RA	Hydrolase family 9	Paper	3.8	<0.001
Lep_00015101-RA	Glucosidase	Paper	3.6	<0.001
Lep_00082758-RA	Glucosidase	Paper	3.4	0.004
Lep_00024218-RA	β-1,6-Glucanase	Paper	2.2	0.04
Switchgrass vs CMC				
Lep_00016310-RA	Hydrolase family 9	CMC	3.6	<0.001

Continued

Table 8.4. Continued

Coding sequence	Enzyme	Up-regulated diet	Log_2-fold change	P_{adj} value
Lep_00016311-RA	Endo-β-1,4-glucanase	CMC	3.9	<0.001
Lep_00016312-RA	Hydrolase family 9	CMC	3.7	<0.001
Lep_00022398-RA	Hydrolase family 9	Switchgrass	2.2	0.002
Lep_00038648-RA	Hydrolase family 9	Switchgrass	3.9	<0.001
Lep_00042851-RA	Hydrolase family 9	Switchgrass	3.3	0.03
Lep_00042852-RA	Hydrolase family 9	Switchgrass	3.9	0.002
Lep_00051683-RA	Hydrolase family 9	Switchgrass	2.2	0.001
Lep_00051684-RA	Hydrolase family 9	Switchgrass	2.2	<0.001
Lep_00052326-RA	Hydrolase family 9	Switchgrass	2.0	0.001
Lep_00071461-RA	Hydrolase family 9	Switchgrass	2.1	0.01
Lep_00095117-RA	Endo-β-1,4-glucanase	Switchgrass	2.8	0.007
Lep_00004873-RA	Glucosidase	Switchgrass	2.0	0.04
Lep_00015101-RA	Glucosidase	Switchgrass	2.8	<0.001
Lep_00019550-RA	Glucosidase	Switchgrass	1.9	0.03
Lep_00020108-RA	Glucosidase	Switchgrass	1.8	0.04
Lep_00021305-RA	Glucosidase	Switchgrass	3.1	0.04
Lep_00024904-RA	Glucosidase	Switchgrass	1.8	0.001
Lep_00036380-RA	Glucosidase	Switchgrass	1.9	0.03
Lep_00048639-RA	Glucosidase	Switchgrass	2.5	0.03
Lep_00052604-RA	Glucosidase	Switchgrass	3.5	<0.001
Lep_00060393-RA	Glucosidase	Switchgrass	2.0	0.01
Lep_00065284-RA	Glucosidase	Switchgrass	3.2	0.003
Lep_00074516-RA	Uncharacterized family 31 glucosidase KIAA1161-like	Switchgrass	2.1	0.001
Lep_00082758-RA	Glucosidase	Switchgrass	2.7	0.04

Continued

Table 8.4. Continued

Coding sequence	Enzyme	Up-regulated diet	Log_2-fold change	P_{adj} value
Lep_00094746-RA	Uncharacterized family 31 glucosidase KIAA1161-like	Switchgrass	1.3	0.03
Lep_00097528-RA	Glucosidase	Switchgrass	2.0	0.04
Lep_00026029-RA	β-Galactosidase	Switchgrass	1.8	0.03
Lep_00028613-RA	β-Glucuronidase-like	Switchgrass	2.8	0.004
Lep_00028614-RA	β-Glucuronidase	Switchgrass	3.1	<0.001
Lep_00046431-RA	β-Glucuronidase	Switchgrass	1.9	0.04
Lep_00053738-RA	β-Glucuronidase	Switchgrass	2.5	0.004
Lep_00065345-RA	β-Glucuronidase	Switchgrass	2.5	0.005
Lep_00034316-RA	β-1,3-Glucan-binding-like	Switchgrass	1.4	0.04
Lep_00087818-RA	β-1,6-Glucanase	Switchgrass	2.6	0.001
Lep_00050000-RA	Hydrolase family 65	Switchgrass	2.9	0.004
Paper vs switchgrass				
Lep_00006775-RA	Hydrolase family 9	Paper	2.8	<0.001
Lep_00014981-RA	Hydrolase family 9	Paper	2.8	0.007
Lep_00014982-RA	β-1,4-Glucanase 5	Paper	3.0	0.002
Lep_00014983-RA	Hydrolase family 9	Paper	3.3	0.03
Lep_00016310-RA	Hydrolase family 9	Paper	3.7	0.003
Lep_00016311-RA	Endo-β1,4-glucanase	Paper	3.7	0.001
Lep_00016312-RA	Hydrolase family 9	Paper	4.1	<0.001
Lep_00036184-RA	Hydrolase family 9	Paper	2.0	<0.001
Lep_00044891-RA	Hydrolase family 9	Paper	2.0	0.003
Lep_00060202-RA	Hydrolase family 9	Paper	2.0	<0.001
Lep_00071132-RA	Hydrolase family 9	Paper	1.4	0.01
Lep_00071461-RA	Hydrolase family 9	Paper	2.1	<0.001

Continued

Table 8.4. Continued

Coding sequence	Enzyme	Up-regulated diet	Log_2-fold change	P_{adj} value
Lep_00078550-RA	Hydrolase family 9	Paper	3.2	<0.001
Lep_00096399-RA	Hydrolase family 9	Paper	4.4	<0.001
Lep_00115213-RA	Hydrolase family 9	Paper	1.8	0.002
Lep_00135466-RA	Hydrolase family 9	Paper	2.0	0.008
Lep_00074516-RA	Uncharacterized family 31 glucosidase KIAA1161-like	Switchgrass	1.3	0.04
Lep_00008899-RA	β-Galactosidase-1 3	Switchgrass	1.5	0.03
Lep_00015051-RA	Mannanase	Switchgrass	2.6	0.007

Table 8.5. Effect of diet on lytic polysaccharide monooxygenase (LPMO) gene expression in rest of the body sample of *Ctenolepisma longicaudata*. Differentially expressed LPMO genes in different pair-wise treatment comparisons of rest of the body sample of *C. longicaudata* fed on carboxymethyl cellulose (CMC), paper, protein, and switchgrass diets ($P_{adj} < 0.05$).

Coding sequence	Up-regulated diet	Log_2-fold change	P_{adj} value
Switchgrass vs protein			
Lep_00006589-RA	Switchgrass	1.9	0.004
Lep_00037556-RA	Switchgrass	1.9	<0.001
Lep_00071838-RA	Switchgrass	2.8	<0.001
Lep_00078707-RA	Switchgrass	1.8	0.01
Lep_00099093-RA	Switchgrass	3.2	0.006
Lep_00127363-RA	Switchgrass	2.7	0.02
Paper vs CMC			
Lep_00018140-RA	Paper	2.1	<0.001
Lep_00030922-RA	Paper	2.3	<0.001
Lep_00030924-RA	Paper	2.1	<0.001
Lep_00037556-RA	Paper	2.9	0.04
Lep_00072794-RA	Paper	2.9	0.02
Lep_00083283-RA	Paper	2.7	0.02
Switchgrass vs CMC			
Lep_00006587-RA	Switchgrass	2.4	0.04
Lep_00006588-RA	Switchgrass	2.7	0.04
Lep_00006589-RA	Switchgrass	2.2	0.02
Lep_00006590-RA	Switchgrass	2.0	0.03
Lep_00018140-RA	Switchgrass	2.0	0.005
Lep_00029394-RA	Switchgrass	2.6	0.04
Lep_00037556-RA	Switchgrass	3.6	0.01
Lep_00039537-RA	Switchgrass	2.2	<0.001
Lep_00043371-RA	Switchgrass	1.6	0.03
Lep_00060573-RA	Switchgrass	2.6	0.03
Lep_00072794-RA	Switchgrass	2.6	<0.001
Lep_00083283-RA	Switchgrass	2.5	<0.001
Lep_00093845-RA	Switchgrass	2.2	0.04
Lep_00099441-RA	Switchgrass	1.6	0.03
Lep_00103234-RA	Switchgrass	2.4	0.001
Lep_00109934-RA	Switchgrass	3.1	0.002
Lep_00119656-RA	Switchgrass	1.8	0.01
Lep_00127363-RA	Switchgrass	3.4	0.02
Paper vs switchgrass			
Lep_00006587-RA	Switchgrass	1.8	0.001
Lep_00006588-RA	Switchgrass	2.5	0.03
Lep_00006589-RA	Switchgrass	1.5	0.02
Lep_00039538-RA	Switchgrass	1.8	0.02
Lep_00127363-RA	Switchgrass	2.6	0.04

commonly expressed in both foregut and rest-of-body samples of *C. longicaudata* independent of diet treatment.

8.3.4 Annotation and alignment analysis of differentially expressed PCWDE genes in *C. longicaudata*

Annotation of all significantly differentially expressed cellulase genes across all feeding treatments and both foregut and rest-of-body in *C. longicaudata* yielded 70 genes encoding for cellulases. The majority (31 sequences) of the differentially expressed cellulase genes encoded for β-1,4-endoglucanases, 30 of which matched to proteins of insect origin and one was most similar to a sea anemone (*Nematostella vectensis*) (Table 8.6). After endoglucanases, the most common differentially expressed cellulolytic enzymes were β-glucosidases (18 genes), 13 of which matched to insect enzymes while three were most similar to molluscs, one was most similar to a fish (*Acanthochromis polyacanthus*), and another was most similar to the crown-of-thorns starfish (*Acanthaster planci*) (Table 8.6). In addition to β1,4-endoglucanases, two sequences encoding for β-1,6-glucanases, which were most similar to molluscan genes, and two β-1,3-glucanases, one similar to a coleopteran (*Dendroctonus ponderosae*) and one most similar to a crustacean (*Daphnia pulex*) gene, were found. Additionally, one mannanase gene was found that was most similar to crustacean (*Daphnia magna*) genes; five α-glucosidase genes were homologous to genes matched to other insect homologs; six β-glucuronidases matching to insect or other arthropod genes; and five β-galactosidases matching to insects were found (Table 8.6). Of the 67 LPMO encoding sequences annotated from the *C. longicaudata* genome, only 15 and 25 LPMO encoding sequences were significantly differentially expressed in foregut and rest-of-body samples, respectively, across different pair-wise treatment comparisons.

The alignment analysis (data not included) of the differentially expressed protein coding sequences of different PCWDEs, paired with their respective insect enzyme class sequences collected from NCBI, showed that 20 out of the 31 differentially expressed genes encoding for β–1,4-endoglucanases in *C. longicaudata* were of full length or near full length and unique from each other. Three of the 11 β–1,4-endoglucanases that were not full length (Lep_00006775-RA, Lep_00036184-RA, and Lep_00071461) had premature stop codons, and the remaining eight were partial sequences. All 18 differentially expressed β-glucosidase sequences were found to be unique, yet only six of them were of full length. The two sequences encoding β–1,6-glucanases and two sequences encoding β–1,3-glucanases were unique, with one full-length sequence in each enzyme group. Out of the five α-glucosidases in the predicted coding sequences dataset, three sequences were of near full length and unique while the rest of them were fragmented. Out of six differentially expressed β-glucuronidase sequences, two were fragmented and duplicated while the rest were unique from each other and fragmented. Additionally, four of five differentially expressed β-galactosidases were found to be unique and complete, while one was fragmented and not aligned well with the rest.

Table 8.6. Differentially expressed plant cell wall degrading enzyme (PCWDE) genes in *Ctenolepisma longicaudata* and their top-blast hit. Significantly differentially expressed PCWDEs ($P_{adj} < 0.05$) were blasted against the NCBInr database and the corresponding protein, glycoside hydrolase (GH) family and tissue (foregut, rest-of-body, or both) in *C. longicaudata* are listed.

Sequence_ID	Identified protein	Organism	Query cover	E-value	% Identity	Accession	GH family	Tissue
Lep_00044891-RA	Uncharacterized protein LOC110840536 (endo-β-1,4-glucanase)	*Zootermopsis nevadensis*	94	0.0	66	XP_021941324.1	GH9	Both
Lep_00115213-RA	Endoglucanase 7-like	*Zootermopsis nevadensis*	84	1e-80	72	XP_021941322.1	GH9	Both
Lep_00071461-RA	Uncharacterized protein LOC110840536 (endo-β-1,4-glucanase)	*Zootermopsis nevadensis*	94	1e-105	72	XP_021941324.1	GH9	Both
Lep_00071754-RA	Endoglucanase E-4	*Blattella germanica*	77	6e-79	60	PSN33998.1	NA	Both
Lep_00016310-RA	Endoglucanase 7-like	*Zootermopsis nevadensis*	84	8e-75	71	XP_021941322.1	GH9	Rest-of-body
Lep_00016311-RA	Cellulase	*Coptotermes acinaciformis*	71	3e-55	58	AAK12339.1	GH9	Rest-of-body
Lep_00016312-RA	Endoglucanase E-4	*Blattella germanica*	38	8e-44	65	PSN33998.1	NA	Rest-of-body
Lep_00052326-RA	Uncharacterized protein LOC110840536 (endo-β-1,4-glucanase)	*Zootermopsis nevadensis*	83	1e-98	50	XP_021941324.1	NA	Both
Lep_00034744-RA	Glycoside hydrolase family 9	*Peruphasma schultei*	86	1e-147	50	AMH40374.1	GH9	Foregut
Lep_00014981-RA	Cellulase	*Antipaluria urichi*	31	4e-58	57	AOV94250.1	GH9	Rest-of-body
Lep_00014982-RA	Putative endo-β-1,4-glucanase of EG1	*Odontotermes formosanus*	82	5e-77	56	BAD12008.1	NA	Rest-of-body

Continued

Table 8.6. Continued

Sequence_ID	Identified protein	Organism	Query cover	E-value	% Identity	Accession	GH family	Tissue
Lep_00014983-RA	Predicted protein (endo-β-1,4-glucanase)	*Nematostella vectensis*	31	7e-23	78	XP_001640311.1	GH9	Rest-of-body
Lep_00071132-RA	Uncharacterized protein LOC110840536 (endo-β-1,4-glucanase)	*Zootermopsis nevadensis*	95	3e-104	72	XP_021941324.1	GH9	Both
Lep_00051683-RA	Glycoside hydrolase family 9	*Timema cristinae*	90	1e-23	64	AMH40395.1	GH9	Both
Lep_00051684-RA	β-1,4-endoglucanase 1	*Panesthia cribrata*	94	8e-36	63	AAF80584.1	NA	Both
Lep_00104311-RA	Hypothetical protein C0J52_21511 (endo-β-1,4-glucanase	*Blatella germanica*	99	2e-55	57	PSN31180.1	NA	Foregut
Lep_00095117-RA	Putative endo-β-1,4-glucanase SmEG1	*Sinocapritermes mushae*	99	1e-47	53	BAD12012.1	NA	Rest-of-body
Lep_00038648-RA	Glycoside hydrolase family 9	*Ramulus artemis*	35	5e-22	65	AMH40383.1	NA	Rest-of-body
Lep_00060202-RA	Endoglucanase E-4	*Blattella germanica*	73	2e-91	62	PSN33998.1	NA	Both
Lep_00022398-RA	Glycoside hydrolase family 9	*Timema cristinae*	88	5e-176	61	AMH40392.1	GH9	Both
Lep_00036184-RA	Uncharacterized protein LOC110840536 (endo-β-1,4-glucanase)	*Zootermopsis nevadensis*	84	0.0	66	XP_021941324.1	GH9	Both
Lep_00034832-RA	β-1,4-endoglucanase 1	*Mastotermes darwiniensis*	70	2e-45	64	AAF63724.1	GH9	Foregut
Lep_00029521-RA	Endo-β-1,4-glucanase	*Panesthia angustipennis*	56	4e-52	51	BAG70031.1	GH9	Foregut
Lep_00078550-RA	Glycoside hydrolase family 9	*Peruphasma schultei*	92	1e-80	59	AMH40374.1	NA	Both

Continued

Table 8.6. Continued

Sequence_ID	Identified protein	Organism	Query cover	E-value	% Identity	Accession	GH family	Tissue
Lep_00135466-RA	Glycoside hydrolase family 9	*Peruphasma schultei*	66	6e-24	73	AMH40374.1	NA	Both
Lep_00096399-RA	Hypothetical protein B7P43_G09674 (Endo-β-1,4-glucanase)	*Cryptotermes secundus*	38	1e-47	62	XP_021924915.1	NA	Both
Lep_00026156-RA	Endoglucanase E-4	*Blattella germanica*	35	2e-70	62	PSN33998.1	GH9	Rest-of-body
Lep_00026154-RA	Putative endo-β-1,4-glucanase HsEG4	*Hodotermopsis sjostedti*	75	8e-43	70	BAD12004.1	NA	Rest-of-body
Lep_00006775-RA	Uncharacterized protein LOC110840536 (Endo-β-1,4-glucanase)	*Zootermopsis nevadensis*	93	0.0	60	XP_021941324.1	GH9	Both
Lep_00042852-RA	β-1,4-endoglucanase	*Mastotermes darwiniensis*	96	2e-107	61	CAD54730.1	GH9	Rest-of-body
Lep_00042851-RA	Cellulase	*Antipaluria urichi*	38	2e-10	49	AOV94250.1	NA	Rest-of-body
Lep_00098705-RA	β-glucosidase	*Acanthaster planci*	25	5e-07	55	XP_022090546.1	NA	Rest-of-body
Lep_00036380-RA	Myrosinase 1-like (β-glucosidase)	*Cryptotermes secundus*	88	2e-104	50	XP_023721112.1	GH1	Rest-of-body
Lep_00004872-RA	β-glucosidase	*Coptotermes formosanus*	76	2e-145	56	AOY34571.1	NA	Rest-of-body
Lep_00004873-RA	Myrosinase 1-like (β-glucosidase)	*Cryptotermes secundus*	94	8e-50	68	XP_023723774.1	NA	Rest-of-body
Lep_00019550-RA	Hypothetical protein B5V51_11090 (β-glucosidase)	*Heliothis virescens*	74	8e-50	34	PCG75727.1	GH1	Rest-of-body

Continued

Table 8.6. Continued

Sequence_ID	Identified protein	Organism	Query cover	E-value	% Identity	Accession	GH family	Tissue
Lep_00021305-RA	Chain A, Crystal Structure of β-Glucosidase from Termite *Neotermes koshunensis* in complex with Tris	*Neotermes koshunensis*	83	3e-92	59	3AHZ_A	NA	Rest-of-body
Lep_00048639-RA	β-glucosidase	*Salganea esakii*	59	7e-24	55	BAO85048.1	NA	Rest-of-body
Lep_00065284-RA	β-glucosidase	*Coptotermes formosanus*	83	1e-60	50	AGM32287.1	GH1	Rest-of-body
Lep_00015101-RA	Myrosinase 1-like (β-glucosidase)	*Cryptotermes secundus*	96	4e-147	50	XP_023721112.1	GH1	Rest-of-body
Lep_00060393-RA	β-glucosidase	*Coptotermes formosanus*	81	5e-55	70	AGM32308.1	GH1	Rest-of-body
Lep_00082758-RA	Lactase-phlorizin hydrolase-like (β-glucosidase)	*Acanthochromis polyacanthus*	90	6e-09	30	XP_022078461.1	NA	Rest-of-body
Lep_00050000-RA	Protein-glucosylgalactosylhydroxylysine glucosidase-like isoform X2	*Spodoptera litura*	90	3e-74	52	XP_022821190.1	NA	Rest-of-body
Lep_00020108-RA	Hypothetical protein LOTGIDRAFT_207250 (β-glucosidase)	*Lottia gigantea*	20	3e-11	83	XP_009064666.1	NA	Rest-of-body
Lep_00052604-RA	Chain A, Crystal Structure of β-glucosidase from Termite *Neotermes koshunensis* in Complex with tris	*Neotermes koshunensis*	89	5e-121	59	3AHZ_A	NA	Rest-of-body
Lep_00024904-RA	Myrosinase 1-like isoform X1 (β-glucosidase)	*Nilaparvata lugens*	82	1e-145	56	XP_022187263.1	GH1	Rest-of-body
Lep_00097528-RA	Lactase-phlorizin hydrolase-like (β-glucosidase)	*Pomacea canaliculata*	46	8e-47	64	XP_025110416.1	NA	Rest-of-body

Continued

Table 8.6. Continued

Sequence_ID	Identified protein	Organism	Query cover	E-value	% Identity	Accession	GH family	Tissue
Lep_00011407-RA	Lactase-phlorizin hydrolase-like (β-glucosidase)	*Pomacea canaliculata*	47	4e-21	47	XP_025110416.1	NA	Rest-of-body
Lep_00113088-RA	Lactase-phlorizin hydrolase-like isoform X2 (β-glucosidase)	*Zootermopsis nevadensis*	97	3e-65	66	XP_021923731.1	GH1	Rest-of-body
Lep_00077538-RA	α-glucosidase family 31	*Periplaneta americana*	89	0.0	53	AIA09350.1	GH31	Rest-of-body
Lep_00074516-RA	Hypothetical protein B7P43_G04891 (Uncharacterized family 31 glucosidase)	*Cryptotermes secundus*	89	1e-147	47	PNF32421.1	GH31	Rest-of-body
Lep_00079073-RA	Hypothetical protein C0J52_14633 (Uncharacterized family 31 glucosidase)	*Blatella germanica*	60	2e-20	65	PSN31308.1	NA	Rest-of-body
Lep_00094746-RA	α-glucosidase family 31	*Periplaneta americana*	83	5e-161	56	AIA09350.1	GH31	Rest-of-body
Lep_00102111-RA	Hypothetical protein L798_13618 (uncharacterized family 31 glucosidase)	*Zootermopsis nevadensis*	67	3e-47	66	KDR11965.1	NA	Rest-of-body
Lep_00065345-RA	β-glucuronidase-like	*Centruroides sculpturatus*	49	2e-11	51	XP_023221887.1	NA	Rest-of-body
Lep_00053738-RA	Hypothetical protein B7P43_G09739 (β-glucuronidase)	*Cryptotermes secundus*	88	1e-67	62	PNF22051.1	NA	Rest-of-body
Lep_00055855-RA	Hypothetical protein B7P43_G09739 (β-glucuronidase)	*Cryptotermes secundus*	96	9e-68	62	PNF22051.1	NA	Foregut
Lep_00046431-RA	β-glucuronidase-like isoform X3	*Cryptotermes secundus*	86	3e-123	57	XP_023718877.1	GH2	Rest-of-body
Lep_00028614-RA	β-glucuronidase-like isoform X3	*Cryptotermes secundus*	91	1e-69	52	XP_023718877.1	NA	Both

Continued

Table 8.6. Continued

Sequence_ID	Identified protein	Organism	Query cover	E-value	% Identity	Accession	GH family	Tissue
Lep_00028613-RA	β-glucuronidase-like	*Halyomorpha halys*	42	6e-46	72	XP_024217164.1	GH2	Rest-of-body
Lep_00059101-RA	β-galactosidase-1-like protein 2	*Cephus cinctus*	74	9e-65	51	XP_015598574.1	NA	Rest-of-body
Lep_00006757-RA	β-galactosidase-1-like protein 2	*Cephus cinctus*	32	4e-62	52	XP_015598574.1	NA	Foregut
Lep_00026029-RA	Hypothetical protein B7P43_G16708 (β-galactosidase)	*Cryptotermes secundus*	66	7e-65	58	PNF39271.1	GH35	Rest-of-body
Lep_00008899-RA	β-galactosidase-1-like protein 2 isoform X2	*Cryptotermes secundus*	95	0.0	43	XP_023717284.1	GH35	Rest-of-body
Lep_00036877-RA	β-galactosidase-1-like protein 2	*Cephus cinctus*	88	3e-60	41	XP_015598574.1	NA	Rest-of-body
Lep_00039163-RA	Hypothetical protein D910_11210 (β-1,3-glucanase)	*Dendroctonus ponderosae*	68	1e-09	44	ERL93924.1	NA	Foregut
Lep_00034316-RA	Hypothetical protein DAPPUDRAFT_203138 (β-1,3-glucanase)	*Daphnia pulex*	49	3e-36	42	EFX69036.1	GH16	Rest-of-body
Lep_00012530-RA	Endo-β-1,4-mannanase	*Daphnia magna*	90	3e-117	54	KZS10752.1	NA	Foregut
Lep_00087818-RA	Hypothetical protein LOTGIDRAFT 101222 (endo-β-1,6-glucanase)	*Lottia gigantea*	70	3e-122	49	XP_009054280.1	GH30	Rest-of-body
Lep_00024218-RA	Hypothetical protein LOTGIDRAFT_101222 (β-1,6-glucanase)	*Lottia gigantea*	72	1e-19	43	XP_009054280.1	NA	Rest-of-body

Among the 67 sequences encoding for LPMOs that were predicted in the *C. longicaudata* genome, 30 sequences were of full length and among these, 29 were unique from each other. In addition, 18 of the LPMO encoding sequences were different from the rest of the *C. longicaudata* LPMOs and reference sequences, suggesting they belong to a new and uncharacterized LPMO family. The remainder of the LPMO encoding sequences were fragmented and most of them did not exactly match to any full-length sequences, indicating that they may be fragments of unique sequences rather than duplicates.

8.4 Discussion

The lack of bias, independence from a requirement for available genomic resources, and high sensitivity presented by whole transcriptome shotgun sequencing support the use of this RNA-seq methodology for examining differential gene expression in non-model organisms. In this report we demonstrate how comparative RNA-seq results were used to identify predicted PCWDE and LPMO genes that were differentially expressed in foregut and rest-of-body tissues in response to feeding restricted to diets containing different types of cellulose. This effort builds upon previous evidence that the foregut contains the highest cellulase activity in Zygentoma (Pothula *et al.*, 2019), by comparing differential gene expression in the foregut tissue from *C. longicaudata* with cellulase activity in the remainder of the insect rest-of-body, which allows for detection of PCWDE and LPMO genes that may be expressed in salivary glands, midgut, and hindgut.

Differential expression of cellulase genes in foregut and rest-of-body samples of *C. longicaudata* fed on protein, paper, CMC, or switchgrass diets was primarily driven by tissue rather than diet (Fig. 8.1). Thus, in most cases cellulase gene expression was localized to a particular tissue (foregut versus rest-of-body) independently of the diet. Almost all of the up-regulated genes in foregut tissue were encoding for endoglucanases, and the majority of up-regulated genes in the rest-of-body sample encoded for β-glucosidases. This observation in *C. longicaudata* supports hypotheses proposed for other insects (Fischer *et al.*, 2013), that cellulose digestion is compartmentalized in insect guts (Fig. 8.2 and Table 8.6). According to the hypothesis, the long cellulose chain is first digested into shorter pieces by endoglucanases that are up-regulated in the foregut tissue, which is followed by further digestion of intermediate products into glucose by β-glucosidases, which were up-regulated in the rest-of-body samples that include midgut and hindgut tissues. However, it is important to note that given the samples analyzed, we cannot conclusively determine the tissue origin of the higher relative expression of β-glucosidases in the rest-of-body sample. Further work quantifying transcript levels of individual tissues would be necessary to provide experimental evidence for compartmentalization of cellulose digestion.

Although cellulase gene expression was primarily driven by tissue type, within each tissue the type of diet presented a regulatory influence on cellulase gene expression. Among all pair-wise treatment comparisons, more

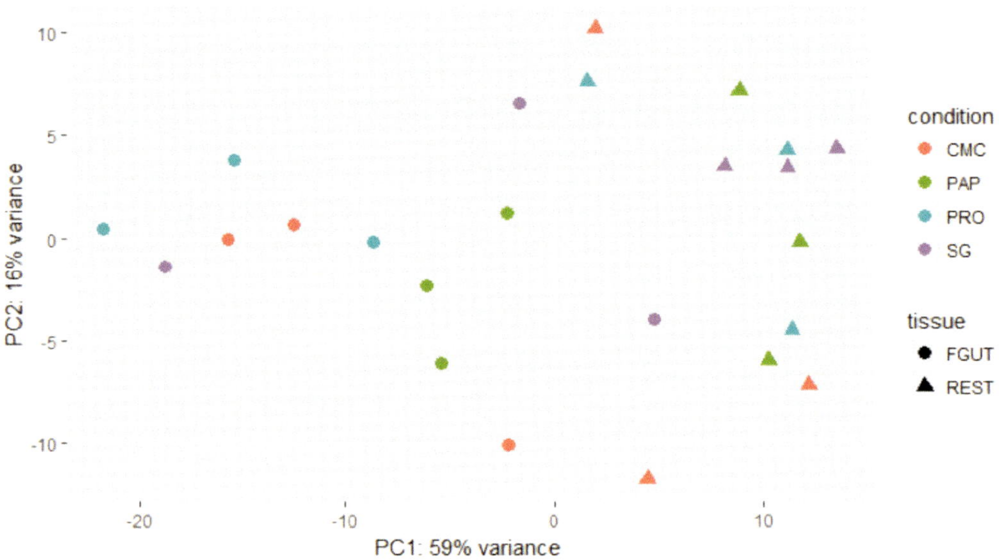

Fig. 8.1. Overall effect of tissue (PC1) and condition (diet, PC2) on gene expression of lytic polysaccharide monooxygenases (LPMOs) in *Ctenolepisma longicaudata* demonstrating a strong expression effect in foregut (FGUT) and rest-of-body (REST) tissues when plotted against gene expression for *C. longicaudata* fed carboxymethyl cellulose (CMC), paper (PAP), protein (PRO) or switchgrass (SG) diets (P_{adj} <0.05).

PCWDEs were up-regulated in the foregut tissue of *C. longicaudata* when fed on paper and compared with the other diets (Table 8.2); and up-regulation of these cellulase genes is occurring in response to the diet that contained the highest cellulose content among the diets tested. A switchgrass diet also resulted in up-regulated expression of a few cellulases compared with a protein diet (Table 8.2). This result may be an outcome associated with the more recalcitrant nature of cellulose that was present in the switchgrass dietary substrate.

Similar to cellulases, differential LPMO gene expression was also mostly driven by the tissue rather than diet, although differences in expression were detected when comparing between some diets (Fig. 8.1). Nevertheless, almost all significantly differentially expressed LPMOs were up-regulated in foregut tissue compared with the rest-of-body (Fig. 8.3). This up-regulation in foregut tissue supports prior observations that the LPMOs enzymes act on long cellulose polymers by forming abrasions or small cuts in fibers and making cellulose more accessible to cellulases (Villares *et al.*, 2017).

Overall, 245 predicted coding sequences were found to be encoding for cellulases in the draft genome of *C. longicaudata*, out of which 70 genes were significantly differentially expressed in *C. longicaudata* across tissues and treatments (Table 8.1). Almost half of the endoglucanases (44.93%) and most of the β-glucosidases (81.82%) that were found in the draft genome responded significantly to cellulose content of the diet. These two enzyme types comprise the main cellulase complex in insects that enable breakdown

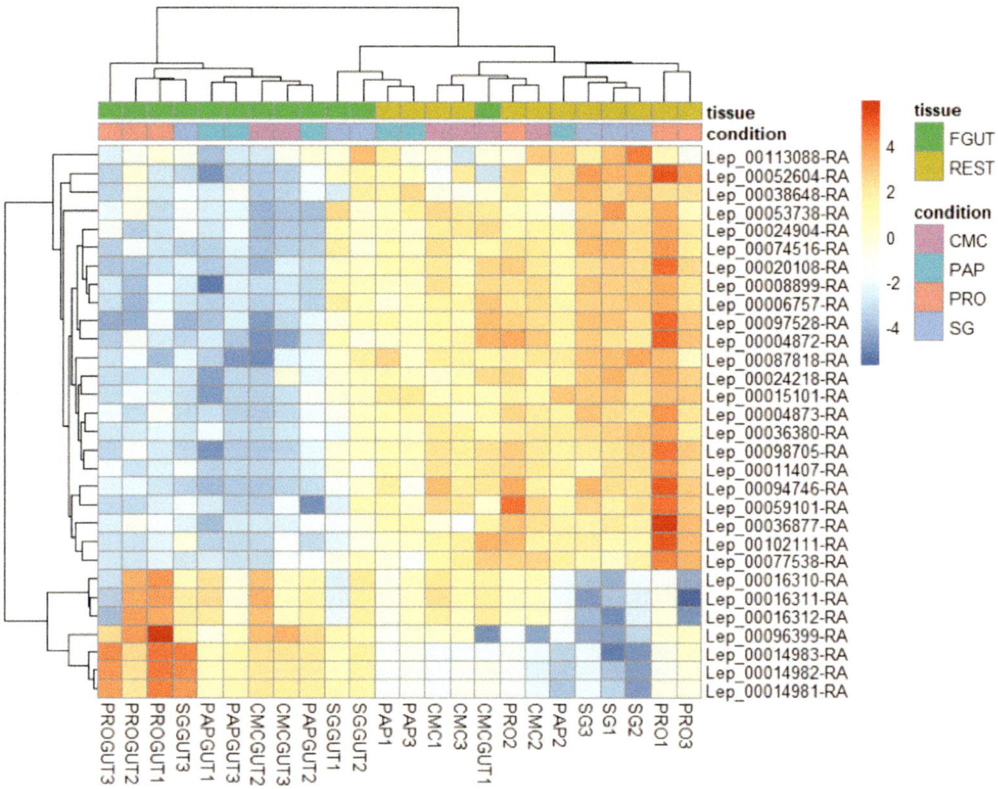

Fig. 8.2. Heatmap of expression of all differentially expressed PCWDE genes in *Ctenolepisma longicaudata* illustrating differences in PCWDE expression in the *C. longicaudata* foregut (FGUT) and rest-of-body (REST) when fed on carboxymethyl cellulose (CMC), paper (PAP), protein (PRO) and switchgrass (SG) diets (P_{adj} <0.05).

of complex cellulose into glucose (Calderón-Cortés *et al.*, 2012). Additionally, out of 67 predicted LPMO encoding sequences in the *C. longicaudata* draft genome, 34 LPMOs were differentially expressed, suggesting that these genes are likely to be involved in more effective cellulolytic digestion. In addition, these genes may possibly help explain why this primitive insect group presents the highest relative cellulase activity in digestive fluids when compared to other insects (Pothula *et al.*, 2019).

Most of the annotated PCWDEs in *C. longicaudata* matched with highest identity to proteins from termites and cockroaches, which may be indicative of close evolutionary relationships between these groups when considering cellulose digestion. In fact, we found PCWDEs in *C. longicaudata* that were most identical to enzymes in primitive hexapod groups such as Collembola (*Orchesella cincta*), or to highly evolved insect groups such as hymenopterans, supporting the conservation of these PCWDEs through insect evolution rather than derived from frequent horizontal transfer from microorganisms (Calderón-Cortés *et al.*, 2012).

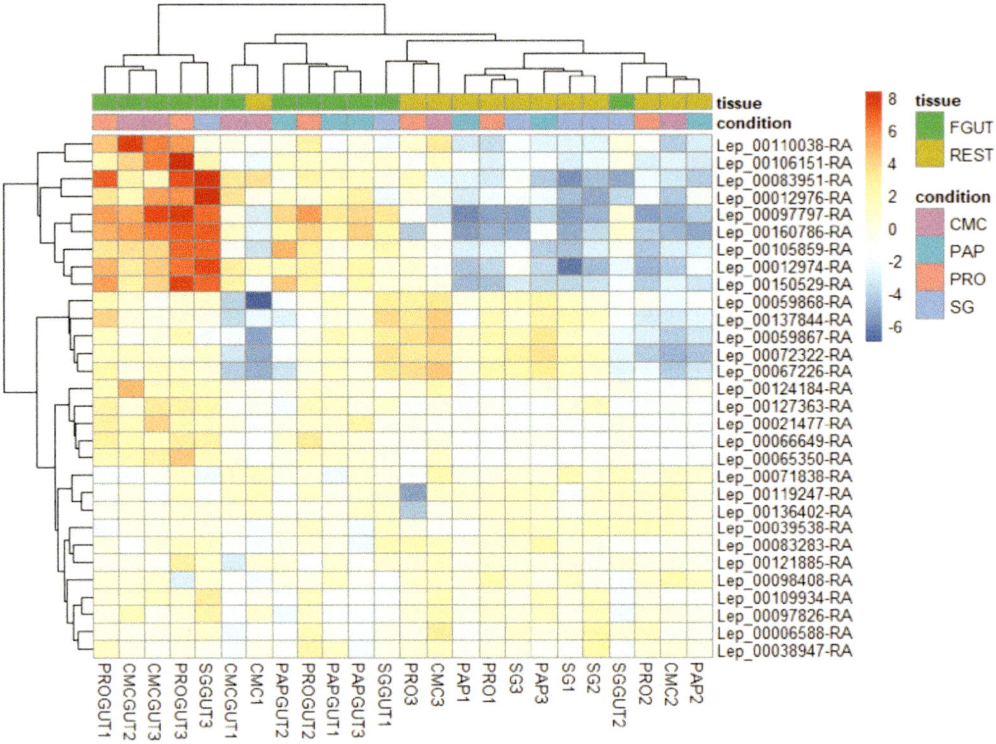

Fig. 8.3. Differential expression of all lytic polysaccharide monooxygenases (LPMOs) in *Ctenolepisma longicaudata* showing differences in LPMO expression in the *C. longicaudata* foregut (FGUT) and rest-of-body (REST) when fed on carboxymethyl cellulose (CMC), paper (PAP), protein (PRO) and switchgrass (SG) diets (P_{adj} <0.05).

Finally, this study demonstrates the utility of RNA-seq methods in identifying genes based on their putative functionality, in this case cellulose digestion, in a primitive insect group with limited genetic resources. Future work will concentrate on functional characterization of the differentially expressed PCWDEs and LPMOs to identify enzymes responsible for high cellulolytic activity amenable to biotechnological applications in improving efficiency of biofuel production.

References

Brand, P., Robertson, H.M., Lin, W., Pothula, R., Klingeman, W.E. *et al.* (2018) The origin of the odorant receptor gene family in insects. *eLife* 7, e38340. DOI: 10.7554/eLife.38340.

Calderón-Cortés, N., Quesada, M., Watanabe, H., Cano-Camacho, H. and Oyama, K. (2012) Endogenous plant cell wall digestion: a key mechanism in insect evolution. *Annual Review of Ecology, Evolution, and Systematics* 43(1), 45–71. DOI: 10.1146/annurev-ecolsys-110411-160312.

Cantarel, B.L., Korf, I., Robb, S.M., Parra, G., Ross, E. *et al.* (2008) MAKER: an easy-to-use annotation pipeline designed for emerging model organism genomes. *Genome Research* 18(1), 188–196. DOI: 10.1101/gr.6743907.

Ekblom, R. and Galindo, J. (2011) Applications of next generation sequencing in molecular ecology of non-model organisms. *Heredity* 107(1), 1–15. DOI: 10.1038/hdy.2010.152.

Fischer, R., Ostafe, R. and Twyman, R.M. (2013) Cellulases from insects. In: Vilcinskas, A. (ed.) *Yellow Biotechnology II. Advances in Biochemical Engineering/Biotechnology*. Springer, Berlin, Heidelberg, pp. 51–64.

Gilbert, H.J. (2010) The biochemistry and structural biology of plant cell wall deconstruction. *Plant Physiology* 153(2), 444–455. DOI: 10.1104/pp.110.156646.

Gomez, L.D., Clare, G.S. and McQueen-Mason, J. (2008) Sustainable liquid biofuels from biomass: the writing's on the walls. *New Phytologist* 178(3), 473–485. DOI: 10.1111/j.1469-8137.2008.02422.x.

Love, M.I., Huber, W. and Anders, S. (2014) Moderated estimation of fold change and dispersion for RNA-seq data with DESeq2. *Genome Biology* 15(12), 550. DOI: 10.1186/s13059-014-0550-8.

Patro, R., Duggal, G., Love, M.I., Irizarry, R.A. and Kingsford, C. (2017) Salmon provides fast and bias-aware quantification of transcript expression. *Nature Methods* 14(4), 417–419. DOI: 10.1038/nmeth.4197.

Pothula, R., Shirley, D., Pereira, O.P., Klingeman, W.E., Oppert, C. *et al.* (2019) The digestive system in Zygentoma as an insect model for high cellulase activity. *PLoS ONE* 14(2), e0212505. DOI: 10.1371/journal.pone.0212505.

Sabbadin, F., Hemsworth, G.R., Ciano, L., Henrissat, B., Dupree, P. *et al.* (2018) An ancient family of lytic polysaccharide monooxygenases with roles in arthropod development and biomass digestion. *Nature Communications* 9(1), 756. DOI: 10.1038/s41467-018-03142-x.

Scrivener, A.M. and Slaytor, M. (1994) Properties of the endogenous cellulase from *Panesthia cribrata* Saussure and purification of major endo-β-1,4-glucanase components. *Insect Biochemistry and Molecular Biology* 24(3), 223–231. DOI: 10.1016/0965-1748(94)90001-9.

Villares, A., Moreau, C., Bennati-Granier, C., Garajova, S., Foucat, L. *et al.* (2017) Lytic polysaccharide monooxygenases disrupt the cellulose fibers structure. *Scientific Reports* 7(1), 40262. DOI: 10.1038/srep40262.

Wang, Z., Gerstein, M. and Snyder, M. (2009) RNA-Seq: a revolutionary tool for transcriptomics. *Nature Review Genetics* 10(1), 57–63. DOI: 10.1038/nrg2484.

Watanabe, H. and Tokuda, G. (2010) Cellulolytic systems in insects. *Annual Review of Entomology* 55(1), 609–632. DOI: 10.1146/annurev-ento-112408-085319.

Zinkler, D. and Götze, M. (1987) Cellulose digestion by the firebrat *Thermobia domestica*. *Comparative Biochemistry Physiology Part B* 88(2), 661–666. DOI: 10.1016/0305-0491(87)90360-9.

9 Using RNA-seq to Help Identify Functions in Unknown Organs

MATAN SHELOMI*

Department of Entomology, National Taiwan University, Taipei, Taiwan

9.1 Introduction

The modern academic environment has increasingly less room for research that is not hypothesis-driven. Observational or descriptive research is viewed pejoratively, with proposals or papers that do not set about to test a hypothesis less likely to be accepted (Casadevall and Fang, 2008). Simply filling in a lacuna of knowledge is not seen as a worthy goal. More pragmatically, an observational study with no guarantee of a falsified or supported hypothesis that can be published is not seen as a solid investment. This strictly Popperian attitude is unfortunate, not only because of the major advances derived from such 'fishing expedition' work, such as the discovery of the Philadelphia translocation associated with leukemia (Rowley, 1973), but also because it ignores the chronological order of science: identify a problem, gather information, draw a hypothesis based on this information, and test the hypothesis. One cannot come up with a hypothesis until one has some preliminary data to start working with, any more than a detective solving a crime can start ruling out suspects before they come up with a list of suspects based on the clues they gathered. The existence of the crime is justification enough to start gathering clues.

Enter high-throughput sequencing. Just as forensic investigators pore over a crime scene and thoroughly search every surface for any possible trace of the perpetrator, not knowing in advance what will be relevant or critical to solving the case, so too do next-generation 'omics' methods allow scientists to examine every gene, transcript, or protein within their samples, whether they will ultimately need them or not. Faced with an open question for which no preliminary data exist to draw hypotheses, next-generation sequencing (NGS) allows for a large amount of data to be generated and mined, with the hope (controlled to some extent by, say, comparing two tissue types or

*mshelomi@ntu.edu.tw

running comparisons across multiple organisms) that patterns will appear leading to a hypothesis that can be more directly tested in the future. In addition, the large molecular datasets produced by next-generation sequencing can reveal questions that we did not know to ask: the 'unknown unknowns,' which could not have been found otherwise, since nobody had known to look for them.

RNA sequencing (RNA-seq), also known as whole-transcriptome shotgun sequencing, is the high-throughput sequencing of all the transcribed RNA in a sample at a single moment in time. Though the researchers may have a specific goal in mind when performing the transcriptome, what they find is not directed by their question. The data will be everything, relevant or not, determined only by the limitations inherent in the methods used, such as sequencing depth or sample preservation quality. Within the treasure trove of data will (hopefully) be, if not the answer to the question, then RNA hints as to where the answer will be found, pending targeted genomic or proteomic studies, which may be in biological spaces the investigators would not have thought to examine. In addition, the data may provide answers to questions other researchers have: this possibility motivates the sharing of NGS datasets on genomics databases and the carrying out of large-scale transcriptomics projects such as 1KITE (the 1000 Insect Transcriptome Evolution Project). Lastly, the dataset may reveal findings so unexpected, so 'unexpectable', that entirely new projects must be drawn up to answer these new questions that have arisen. This high potential for discovery is one of the strongest selling points of RNA-seq, which I shall now illustrate by example: the mysteries of the Phasmatodean midgut.

9.2 Phasmatodea Digestion and Anatomy

9.2.1 Prior work

The Phasmatodea, or leaf and stick insects, are an order within the Polyneoptera closely related to the Embioptera (webspinners) (Letsch *et al.*, 2012). Though iconic in appearance, the order is small and its members are not major pests and do not vector diseases or even bite, and so Phasmatodea is not as thoroughly studied as other clades. Their diet is almost exclusively herbivorous, without any reported carnivory, detritivory, or parasitism (Calderón-Cortés *et al.*, 2012). Outside of scattered reports from breeders, the literature suggests that leaves are their only source of food. The question that arises from this observation is: how do phasmids extract the necessary nutrients from a leafy diet? While not as limiting as the xylophageous diet of termites, a leaf diet still has large amounts of cellulose to contend with in the cell walls surrounding the nutritious cytoplasm within. In addition to cellulose, other plant cell wall component polysaccharides include hemicellulose (xylan, xyloglucan), pectin (polygalactose), and lignin.

The anatomy of a herbivore's digestive system can provide hints as to how it extracts nutrients from recalcitrant plant matter. Whereas carnivore guts tend to be short and unadorned, in herbivores one finds multiple twists and turns to increase digestion time, as well as compartmentalization and enlargements or paunches that serve as fermentation chambers in which cellulose and other polysaccharides are broken down. The relevant plant cell wall degrading enzymes (PCWDEs) can be endogenously produced by the herbivore itself (in which case genes for these enzymes will appear in genomic or transcriptomics studies), or from symbiotic microbes (Calderón-Cortés *et al.*, 2012). In the latter, the digestive system will be modified to house these microbes in specialized cells (bacteriocytes or mycetocytes) or chambers in which the undigested matter is exposed to bacterial enzymes for long enough to be maximally degraded. The former was once thought to be rare, yet the current thinking is that endogenous animal PCWDEs are the rule rather than the exception: the ancestral Metazoan is believed to have had a cellulase gene (Davison and Blaxter, 2005), and the absence of these enzymes in many heterometabolous insects (Diptera, Lepidoptera, and most Hymenoptera and Coleoptera so far) is considered to have been a loss of function rather than the ancestral insect state as once believed (Fischer *et al.*, 2013).

What does the Phasmatodea digestive system say? Little. The gut (Fig. 9.1), constrained by the long and thin bauplan of the mimetic order, is long and thin, with no noticeable fermentation chamber and no convolutions, and only rudimentary gastric caecae (Shelomi *et al.*, 2015). The gut can be roughly divided as follows: a crop, normally filled with still discernible chunks of green leaves; a proventriculus partially obscured by the gastric caecae that is internally ridged and lined with small spines and which reduces the leaves into a brown slurry; the midgut, where digestion occurs; and the hindgut, where water is resorbed from the frass (Shelomi *et al.*, 2015). At the junction of the midgut and hindgut are Malpighian tubules as in other insects. No bacteriocytes or mycetocytes are visible here or in the fat bodies or salivary glands of the Phasmatodea: histology, culturing, and metagenomics have all concluded that Phasmatodea digestion is microbe

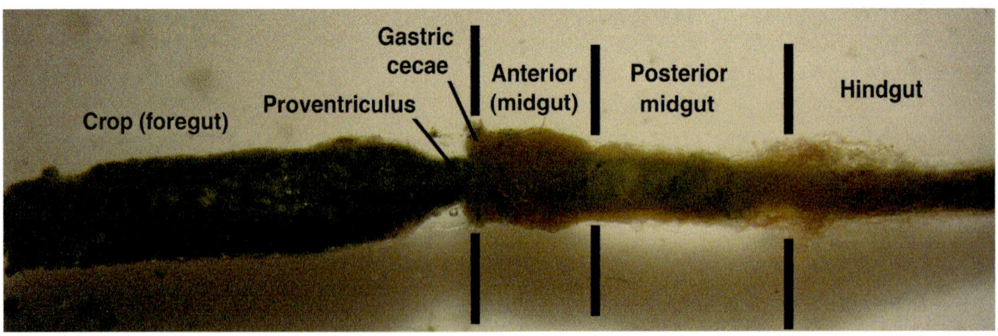

Fig. 9.1. Digestive tract of the Phasmatodea (*Areaton asperrimus*). Different sections of the gut are labeled. The insect had been vitally stained with May Grünwald stain for 30 min prior to dissection. The Malpighian tubules arising from the junction between the midgut and hindgut are visible, slightly pigmented. Photo credit: Matan Shelomi.

independent (Shelomi *et al.*, 2013, 2015). If PWCDEs are present, then they must be endogenous.

However, the midgut of the Phasmatodea is unique among the Insecta. The anterior portion is easily distinguished from the rest, with thick, accordion-like pleats perpendicular to the body length that greatly increase the midgut's internal surface area. The folding stops midway into the midgut, after which appear a variable number of small, pyriform ampules (Fig. 9.2) from which long, thin tubes emerge and corkscrew down towards the hindgut, intermingling among the Malpighian tubules, which are about three times the thickness. The ampules are hollow, opening into the midgut lumen. The tubules are motile, tracheated, and hollow, with their contents flowing anteriorly towards the ampule before emptying into the midgut (Shelomi *et al.*, 2015). These ampules and their thin tubules are a Phasmatodea-specific synapomorphy, not found in any other organism (Bradler, 2015). The earliest description may be from 1896, where they were hypothesized to be excretory organs like the Malpighian tubules (Bordas, 1896). Numbering anywhere from a pair or two in the basal Timematidae to several dozen in other species (Shelomi *et al.*, 2015), they have been called 'appendices of the midgut' for at least a century (De Sinéty, 1901) and have had no known function for about as long (Savage, 1962). Most authors guessed that they are excretory (Cameron,

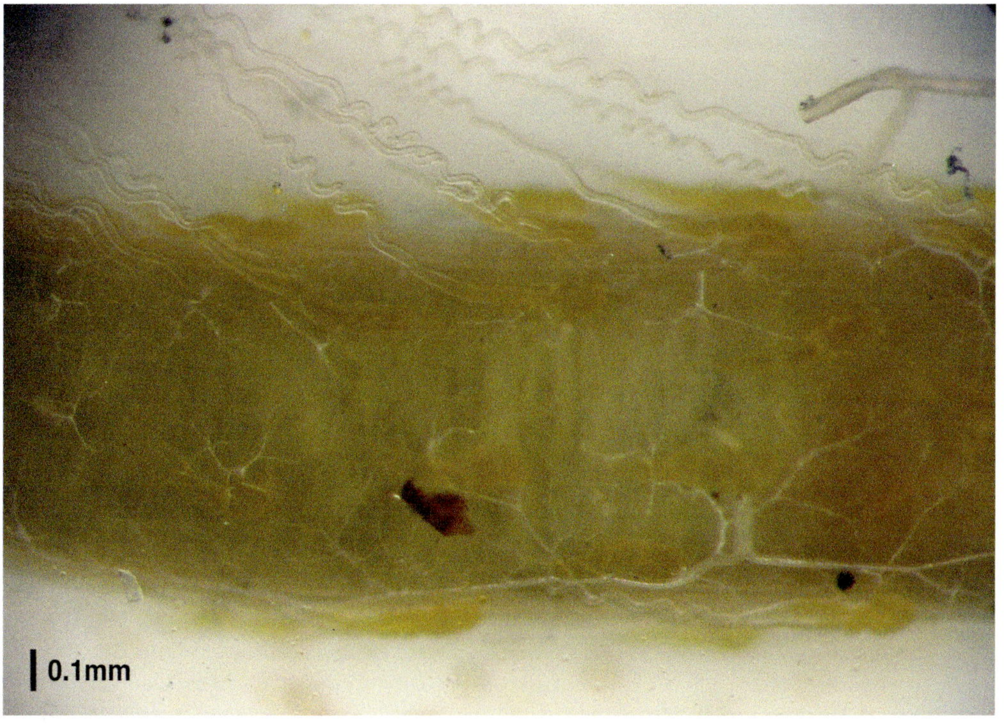

Fig. 9.2. Close-up of the posterior midgut of a Phasmatodea (*Ramulus artemis*). The yellow ampules of the midgut appendices and the clear tubules that run from them towards the posterior of the insect are clearly visible. Photo credit: Matan Shelomi.

1912), but until the 21st century none ever demonstrated this: a striking omission considering that much of what we know about the Malpighian tubules was discovered in Phasmatodea themselves, namely in the work of Ramsay, who was aware of the appendices and developed several methods used to study Malpighian tubule function, but for whatever reason never employed the latter to the former (Ramsay, 1954, 1955). Following a paper in 1962 they were ignored by the scientific community for half a century, when a microbiological study of the anterior and posterior midgut demonstrated that, whatever the function of the appendices is, housing symbiotic microbes is not one of them (Shelomi *et al.*, 2013).

So what do the appendices do? Experiments with vital staining confirmed that they are excretory, but selecting stains differently from how other excretory organs such as the Malpighian tubules or pericardial tissue do (Shelomi and Kimsey, 2014). The posterior midgut is more alkaline than the anterior midgut, with alkalinization beginning at the origin of the appendices (Shelomi *et al.*, 2015), and authors confirmed chemically that carbonic anhydrase is most active in the appendices, producing bicarbonate and raising the pH of the midgut lumen (Monteiro *et al.*, 2014). A high pH may help in digesting hemicellulose, but so would the high surface area of the acidic anterior midgut. What was happening in the gut on a molecular level? Did Phasmatodea produce PCWDEs or not? And what role did the midgut appendices have in this?

9.2.2 Search for endogenous PCWDEs

The earliest work into the digestive enzymes of Phasmatodea was my own proteomics-based search for cellulases (Shelomi *et al.*, 2014b). We found high cellulase activity in the anterior midgut and two genes of β1,4-endoglucanase (cellulase) of the glycoside hydrolase family 9 (GH9) – the same type thought to have been in the ancestor to all Metazoa – in the stick insects *Eurycantha calcatara* and *Entoria okinawaensis*. The sequences were similar to those of termite endocellulases, suggesting that the enzymes were endogenous and primarily expressed in the pleated and folded anterior midgut. Yet unknowns remained: were these the only enzymes active in the gut? Are the cellulases expressed by the anterior midgut tissue itself, or are they produced elsewhere, such as in the gastric caecae or the appendices, and then brought to the anterior midgut via countercurrent flow (Monteiro *et al.*, 2014)? Reviewers too had doubts about the endogenicity of these enzymes: the prevailing but already disproven thought that cellulases are only expressed in microbes and possibly termites was still strong in the early 21st century (if not today), despite the absence of evidence for microbial origin of these newly discovered phasmid enzymes and the amino acid and nucleotide sequence homologies and antigency against an endogenous termite cellulase anti-serum all clearly pointing towards endogenicity (Shelomi *et al.*, 2014b).

To answer the open questions and appease the critics, I undertook a *de novo* transcriptomics project (Shelomi *et al.*, 2014a). Using transcriptomes to study digestive enzymes was not new. It has always had the potential to identify new digestive enzymes, among other biologically important proteins such as cytochrome P450 mono-oxygenases and glutathione-S-transferases involved in detoxification (Pauchet *et al.*, 2010b). Transcriptomics using pyrosequencing of beetles had previously revealed nearly 200 PCWDEs from eight different enzyme families, demonstrating that the prior absence of PCWDEs from model Coleopterans had been a sampling artifact. The enzymes found included endoglucanases of the GH families 45 and 48, rather than the GH9 found in termites and phasmids, as well as polygalacturonase or pectinase, GH28 (Pauchet *et al.*, 2010a). Without knowing what we would find, other than the expectation of a cellulase or two, we set up the first transcriptomics study of the Phasmatodea (Shelomi *et al.*, 2014a).

Because genomic resources for the non-model Phasmatodea were not available, other than a draft genome for the basal and unique genus *Timema* of the family Timematidae that has since been published (Soria-Carrasco *et al.*, 2014), we used the Trinity method to recover as many full-length transcripts without aligning reads to the non-existent reference genomes (Haas *et al.*, 2013). Six species from several Phasmatodea families were chosen: *Peruphasma schultei*, *Sipyloidea sipylus*, *Aretaon asperrimus*, *Extatosoma tiaratum*, *Medauroidea extradentata*, and *Ramulus artemis*. From five individuals in each species, we made separate RNA libraries for the anterior midgut, where cellulase expression was hypothesized to occur, and the posterior midgut, which housed the enigmatic appendices. Tissues were swiftly dissected under sterile conditions from surface-sterilized individuals, with the gut contents and any microbes on them removed through several rinses in 70% ethanol, followed by homogenization using the TRIzol® Reagent protocol. RNA was extracted and RNA libraries produced for sequencing on an Illumina HiSeq platform. For *Peruphasma shultei*, three biological replicates of five individuals each were made, and whole transcriptome shotgun sequencing, or RNA-seq, performed using RSEM (Li and Dewey, 2011) to calculate gene expression levels and the R package EBSeq (Leng *et al.*, 2013) to compare these across the two tissues. This allows one to quantify which transcripts are highly expressed in each tissue type, and which ones are differentially up-regulated or down-regulated between the two, without the need for knowing *a priori* what genes are in the system as in microarrays. While studying the digestive enzymes was the official justification for this project, the possibility that this research could shed light on the mystery of the appendices was not out of mind at this time.

To annotate the transcriptome, the program Blast2GO is effective and produces excellent figures, but slow if one is using the NCBI translated nucleotide collection database, which we must for a *de novo* transcriptome. Thus, Blast2GO was limited to the RNA-seq data and to the top 500 most highly expressed sequences from the other five species. To specifically find PCWDEs, we ran a BLAST search using a query of known endogenous insect, fungal, bacterial, and protozoan PCWDEs. To make doubly sure any enzymes were endogenous, putative PCWDEs were checked for signal peptides and poly-A tails. To make triply sure any enzymes were endogenous, a genomic DNA library was produced from the microbe-free *Peruphasma schultei* brain, from which homologs to any endogenous PCWDEs from the gut samples should be found. Every PCWDE to be discussed in this chapter passed all of these tests for endogenicity, though a few reviewers still found it hard to believe that non-microbial cellulases exist, so deep in the contemporary biologist's consciousness runs this misconception.

The results of this study (Shelomi *et al.*, 2014a), which at the time was the biggest Phasmatodea transcriptomics project to be done, included expected, unexpected, and unexpectable results. Endogenous GH9 cellulases were found, expressed by the anterior midgut tissue as hypothesized, but in large numbers: while most insects known to produce cellulases had one or two genes, the Phasmatodea species had as many as seven isogroups each. Why would they need so many? Cellobiases (beta-glucosidases) were found, suggesting Phasmatodea may have the biological means to degrade cellulose completely into glucose monomers. Were they doing this? Each phasmid produced up to three β1,3-glucanases, which are a little-known polysaccharidase thought to play more of a role in breaking down microbial cell walls than in digestion (Pauchet *et al.*, 2009).

The most surprising find of all, however, was the presence of polygalacturonase, or pectinase. Again, multiple pectinase genes were expressed, up to 21 per species. This was a very unusual find, as pectinases really were not supposed to be endogenously expressed in insects. Indeed, the sequences of these pectinases matched those of gammaproteobacteria rather than insects ... but they were unquestionably endogenous. Despite the bacteria-like sequence homology, the genes all had eukaryote-specific signal peptides, considerable polyadenylation, and homologs within the brain genome. The only test they failed was being found in the *Timema* draft genome: they were absent. We hypothesized that these genes had been acquired through horizontal gene transfer from a bacterium at some point in the Phasmatodean evolutionary history. This, too, is not unusual: horizontally acquired fungal pectinase genes were known to be endogenously expressed in Hemiptera and certain Coleoptera (Kirsch *et al.*, 2014; Shen *et al.*, 2003). Challenging yet another deeply believed myth, inter-domain horizontal gene transfer appears to be far more common than previously thought: a rule rather than an exception (Boto, 2014). Nonetheless, this was unexpected: nobody could have predicted that Phasmatodea express pectinases, let alone nearly two dozen of them. Only the wide and untargeted net of transcriptomics could have revealed this finding.

Thus, the project generated more questions than it answered, and, alas, brought us no closer to solving the riddle of the midgut appendices. We needed to go deeper.

9.2.3 Evolution of Phasmatodea PCWDEs

The question of the appendices was briefly put on hold to understand what was happening with the cellulases and pectinases. Using the transcriptome data, which was often truncated at the 5′ or 3′ end or both, one could generate primers for the transcripts. Using rapid amplification of cDNA ends

(RACE) PCR, one can reverse transcribe cDNA from the RNA, and then use the primers to sequence the missing ends. Once the complete open reading frame (ORF) for these target genes is available, proper forward and reverse primers for these enzymes can be designed, and used to amplify the relevant portion of the cDNA. This DNA fragment can be heterologously expressed in a vector system such as *Spodoptera frugiperda* cells (*Sf*9 cells), which produce the enzyme into culture medium that can be harvested and used for testing against various substrates.

Briefly, the cellulases of Phasmatodea are not only active against cellulose, degrading it down to the trimer and occasionally dimer level, but also some are also bifunctional and capable of true polymer hydrolase activity against the polysaccharide backbone of xylan, and others similarly capable of hydrolysing xyloglucan (Shelomi *et al.*, 2016b), all while maintaining their cellulolytic abilities. True β-endo-1,4-xylanase activity in an animal gene is uncommon enough, but a bifunctional cellulase–xylanase had never been discovered in any Metazoan before: among GH9 enzymes it was only known from rice. True xyloglucanase activity had never been described in a Metazoan before, and a bifunctional cellulase–xyloglucanase in the GH9 family was unheard of in any life form. What started as a search for cellulases had, entirely unexpectedly, uncovered enzymes that completely rewrote what endoglucanases are supposed to do. Even the more imaginative enzymologists who predicted the existence of bifunctional endoglucanases never expected them to be discovered in animals (Eklöf *et al.*, 2012). As for the pectinases, some were active against citrus pectin and/or polygalacturonic acid, but many were inactive. Whether these enzymes have a non-pectinolytic function or simply required a co-factor or condition not provided in the assay is unknown (Shelomi *et al.*, 2016a).

Returning to transcriptomics, what remained unknown was whether insects other than the Phasmatodea had these enzymes. *Timema*'s lack of pectinases needed to be explained, as did the ubiquity of the bifunctional cellulases: just how wrong was the prevailing dogma about insect digestive enzyme diversity? Rather than generate hundreds of *de novo* transcriptomes, at this time we turned to a source that did the hard work for us: the 1000 Insect Transcriptome Evolution (1KITE) Project. With the stated goal of producing transcriptomes for over 1000 insects, 1KITE has already published some papers based on the 100 or so transcriptomes they have released to the public on the NCBI Expressed Sequence Tag database (Letsch *et al.*, 2012). By contacting the relevant people at 1KITE (http://www.1kite.org/), we were able to have access to several of the completed but not yet publicly available transcriptomes: those of all their Phasmatodea (including *Timema christinae*), and representatives from several Polyneoptera outgroups from the orders Embiidina, Mantodea, Blattodea (including Isoptera, the termites now subsumed within Blattodea), Notoptera/Xenonomia (the Mantophasmatodea and Grylloblattodea), Dermaptera, Zoraptera, Plecoptera, and Orthoptera; as well as their already published transcriptomes that covered much of insect diversity. As we did for the original five *de novo* phasmid transcriptomes, we mined these transcriptomes for PCWDEs of all kinds, including the full genes identified in the RACE PCR study in the query. Thus, we could sample the breadth of insect diversity for the enzymes discovered in Phasmatodea.

The results of the cellulase study, which also included RACE PCR and subsequent heterologous expression and chemical activity tests on enzymes from several species of roach, termite, webspinner, and mantis, are pending publication. Briefly, the gene duplication that led to Phasmatodea having multiple cellulase groups happened early in the evolution of the entire Polyneoptera clade. The bifunctional cellulase–xylanases may be unique to Phasmatodea, while cellulase–xyloglucanases may have evolved twice: once in the Phasmatodea/Embioptera lineage and once in the Blattodea (Shelomi *et al.*, 2019, unpublished results). The results of the pectinase study resolved the question of the horizontal gene transfer of pectinase: it occurred somewhere after the Timematidae (comprising the single genus *Timema*) split from the rest of the Phasmatodea, the Euphasmatodea. The pectinases are missing in *Timema* and in all other insects, and only appear in Euphasmatodea. This is estimated to have occurred sometime between 125 to 103 million years ago. As all the Euphasmatodea have the large number of pectinases, the horizontal gene transfer must have happened after the suborder radiated into the 3000 or so described species today, approximately 61 million years ago (Bradler *et al.*, 2015). Thus, we can time the horizontal gene transfer event from the microbe – likely a gut transient or, possibly, a symbiont since abandoned – to this 60–40 million-year window, though why the one or possibly two genes that transferred subsequently duplicated into two dozen pectinase genes remains a mystery. Note that not all of the 1KITE Euphasmatodea transcriptomes had pectinase genes. However, these aberrant datasets were also missing cellulase enzymes. Upon investigation, the 1KITE scientists involved with Phasmatodea sequencing conceded that they only used the heads and thoraxes of the larger Phasmatodea due to these insects' considerable size. Because the Phasmatodea midgut starts around the junction of the thorax and abdomen (Shelomi *et al.*, 2015), some of the 1KITE samples contained a bit of anterior midgut tissue and others did not. These can be respectively identified by the presence or absence of cellulase enzyme transcripts. The full body of *Timema*, being less than 3 cm long, was used for its transcriptome source, and indeed we found all the right cellulase enzymes in the 1KITE dataset, so we are confident that the absence of polygalacturonase here was not an artifact and that this genus really does lack pectinases. Unfortunately, because the 1KITE Phasmatodea transcriptomes lack transcripts from any abdominal tissues, including the excretory and reproductive systems, one must take any other 1KITE conclusions about Phasmatodea with a serving of skepticism.

9.3 Phasmatodea Midgut Appendices

Returning to the appendices, we now come to the main theme of this chapter: using targeted transcriptomics to identify the function of an organ that, for over a century, was completely unstudied.

The assumptions were that the midgut appendices were excretory in nature (Bordas, 1896; Heymons, 1897; De Sinéty, 1901; Savage, 1962). As we see from the case of animal cellulases, assumptions often do not age well. Yet experimental evidence with vital staining – the injection of dyes into a living organism and subsequent observation of which tissues pick up or eliminate the dye, which is an old but effective method of studying excretory organ function (Lison, 1937, 1942) – confirmed that the appendices do pick up certain xenobiotic compounds from the hemolymph and transport these into the midgut lumen for subsequent elimination in the frass, similar to how Malpighian tubules eliminate salts, nitrogenous wastes, and other vital stains. The transport of some of the vital stains may be an effect of normal metabolic waste elimination. If Malpighian tubules and the midgut appendices transport different substances, the complement of transport genes (cationic, anionic, sugar, calcium, amino acid, sulfate, etc.) should differ between them in accordance with what they transport. On the other hand, the genes responsible for this elimination of xenobiotics may be those involved in pesticide resistance, such as ATP-binding cassette transporters. In turn, any organ involved in xenobiotic elimination might also express genes for xenobiotic metabolism and resistance, such as the aforementioned cytochrome P450s, glutathione-S-transferases, carboxylesterases, and acetyl cholinesterases (Mamidala *et al.*, 2012). The Malpighian tubules express all of these. They are not only excretory organs, but also insecticide resistance organs capable of eliminating xenobiotic toxins (Dow, 2009). Potentially the appendices expressed them as well. All of these gene families, however, are extremely large: the number of transcripts an individual species is estimated to have for these is both high and variable. Once again, the wide but deep net of transcriptomics is needed to resolve the question.

Excretion was not the only active hypothesis, however. Recall that the appendices were found to produce carbonic anhydrase. A safe hypothesis would be that carbonic anhydrase transcripts should be differentially expressed in the appendices compared with other tissues. Still, could they also differentially express other related genes? In dissections, the ampules sometimes contained unidentifiable concretions, and the vital stains they transported were, on average, slightly more cationic than those they did not. Consider also the fact that the Phasmatodea Malpighian tubules are themselves synapomorphically differentiated into two types: primary tubules that eliminate wastes, and secondary tubules that sequester calcium for deposition on the eggs as a Phasmatodea-specific layer of calcium oxalate (Pantel, 1919; Savage, 1962; Hinton, 1981). (Yes, the Phasmatodea are full of unique surprises.) Thus, another hypothesis to be tested is that the appendices are involved in some aspect of cation regulation, calcium transport, biomineralization, etc. Without knowing what specific proteins to look for, again, transcriptomics is needed.

The original *de novo* data did not reveal any meaningful conclusions about the midgut tubules' functions: it is not clear from the differentially expressed transcripts which are expressed by these organs themselves and which are expressed by the posterior midgut wall itself (Shelomi *et al.*, 2014a). More refined tissue-specific transcript profiling was warranted (Shelomi, 2017). This time we used *Carausius morosus*, the Indian walking stick, also called the laboratory walking stick due to being the only Phasmatodea whose widespread use almost makes it a model organism. Indeed, it was on *C. morosus* that Ramsay made his first discoveries into Malpighian tubule function (Ramsay, 1955). Thus, we hoped to close the century-old loop on Phasmatodea excretory organ research in the same species where it started. The tissues to be studied this time were the Malpighian tubules, the appendices of the midgut, and the posterior midgut wall. We needed to include the latter because separating the ampules and tubules of the appendices themselves from the midgut proved to be impossible. However, an unadorned portion of midgut wall exists between the part studded with ampules and the origin of the hindgut and Malpighian tubules. This 'post-posterior midgut' section would serve as the control to the appendices section.

Tissues were quickly dissected on ice from adult female *C. morosus* (the species is parthenogenetic), carefully separated and rinsed of any gut contents or stray tubule snips from the wrong organ, and macerated in RNAlater™ in a frozen-cold bead-beating homogenizer prior to RNA extraction and RNA library production, Illumina HiSeq sequencing, raw data processing, quality filtering, and de-multiplexing. Unfortunately, an annotated genome for *C. morosus* did not yet exist, nor did we feel *Timema* was close enough to *Carausius* to be used as a reference, so once again we had to produce a *de novo* transcriptome from these raw reads, this time using CLC Genomics Workbench v8.0 (Qiagen®) for transcriptome assembly, CAP3 for contig fusion (Huang and Madan, 1999), the Cufflinks v2.2.1 suite of tools (http://cole-trapnell-lab.github.io/cufflinks/) for RNA-seq between the three tissues, and Blast2GO for annotations. In addition to BLAST, we used the program nhmmer to mine the transcriptomes for specific gene types (e.g. ABC transporters) using as a query back-translated protein sequences from relevant insect sequences on the PFAM database (Bateman *et al.*, 2004). At this point, the endogenicity of the resulting transcripts was no longer questioned, and the close sequence homology of the genes identified here to those in other Polyneoptera suggests these are all endogenous *Carausius* genes.

As by now should be guessed, the results included findings we expected, findings we did not expect, and findings we could never have expected (Shelomi, 2017). Several ABC transporters were differentially expressed in the appendices, particularly multi-drug resistance proteins from ABC transporter family C. Many cytochrome P450s and two glutathione-S-transferases were highly and differentially expressed in the appendices. The Malpighian tubules also expressed a large set of ABC transporters and cytochrome P450s, but these were different from what the appendices expressed. The transcriptomics data confirmed what had been suggested by the vital staining: the appendices are involved in xenobiotic resistance, eliminating potentially harmful foreign substances from the hemolymph and, importantly, targeting different molecules than the Malpighian tubules. What was rejected, on the other hand, was the century-old hypothesis that the appendices are involved in excretion-based homeostasis. Ion transporters were abundantly expressed in Malpighian tubules but not the appendices, where only some amino acid and protein transporters and V-ATPase domain-containing transporters that may or may not be involved in xenobiotic transports were up-regulated. In other words, the appendices do not overlap with the Malpighian tubules in function. If the two organs were homologous in origin, they have thoroughly diverged in function.

Carbonic anhydrase was highly expressed in the appendices as predicted, but cationic transporters and other biomineralization genes were not. These instead were up-regulated in the Malpighian tubules, as expected given their role in calcium sequestration in Phasmatodea. The hypothesis that the appendices are involved in biomineralization seems disproven, along with the cation excretion hypothesis suggested by the vital staining. Their alkalinizing effect on the posterior midgut lumen is primarily due to carbonic anhydrase production. However, the single the most strongly up-regulated transcript in the appendices relative to the Malpighian tubules was for a biomineralization gene: otopetrin, a vertebrate calcium homeostasis gene also involved in diverse functions in other organisms, such as firefly light production, mosquito response to juvenile hormone, and a signal pathway in mouse adipose tissue (Hughes *et al.*, 2008; Wang *et al.*, 2014). It is this latter function that might be important in the appendices, due to the final and most remarkable finding.

The top 25 most highly expressed (which we defined as having at least ten times more reads per kilobase million than the average for that tissue) and most differentially expressed (based on $\log_2$ fold change values compared with the other tissues) transcripts in the appendices included several cytochrome P450s, both the glutathione-S-transferases, some transporters, and some metalloproteases of the kind one finds in all cells, as well as two lipases and three fatty acid binding proteins (Shelomi, 2017). When looking at all the fat-related genes (lipid storage droplets surface-binding proteins, fatty acid-binding proteins, lipases and phospholipases of all kinds, etc.), one found that many of them were differentially expressed in the appendices, and often among the most highly expressed genes within this tissue as well. Cholesterol and lipoprotein specific ABC transporters were overexpressed here as well. Some phosphatases were also expressed in the Malpighian tubules, likely due to Phasmatodea excreting excess calcium as a phosphate rather than as calcium carbonate, but the abundance of fat, lipid, and lipophilic substance-related genes were up-regulated in the appendices. These findings were surprising: nothing in the observational data on these organs for the past century ever hinted that they played a role in fat metabolism and/or transport: the ampules are naturally filled with a yellow, non-polar substance, but they can carry water-soluble stains too (Shelomi and Kimsey, 2014). The transcriptome presents another hint as to what is happening: no lipophorin genes were found in the transcriptome, despite this being the main lipid transporter in insects (Canavoso *et al.*, 2001). A scan of the 1KITE transcriptomes did turn up some lipophorin-like genes in some Phasmatodea, but never more than one or two transcripts, compared with several in all other insect transcriptomes. The possibility exists that our *Carausius* transcriptome simply missed the tissues that express lipophorins, but at this stage we cannot rule out that Phasmatodea lack or are deficient in this key fat-transporting molecule and use an alternative system involving the appendices. More research is needed, but now, for the first time, we have a hypothesis to look into that is not just excretion. The appendices transport fatty acids and such compounds: an exaptation that could never have been discovered without transcriptomics any more than the endogenous Phasmatodea pectinases could have.

9.4 Conclusion

To summarize, the appendices of the Phasmatodea midgut are not simply extra excretory organs. They are involved in multi-drug resistance, eliminating foreign compounds from the hemolymph through an array of ABC transporters and cytochrome P450s. They express carbonic anhydrase that alkalinizes the midgut lumen, though which may have useful effects on the functioning of the appendices themselves as well. Lastly and thoroughly unexpectedly, they are important organs of fat metabolism and homeostasis. How exactly they do this is unknown: knowledge on fat metabolism in insects is relatively low. Importantly, this lipid-transport function could never have been discovered in any other way, not because biochemical tests for lipid transport are not possible, but because nobody would have ever thought to apply them to the appendices, similar to how nobody studying animal cellulases ever thought to test for bifunctionality against xyloglucan until we did. The Phasmatodea's ever-increasing collection of unique physiological features shows just how much in biology we do not know that we do not know.

Transcriptomics work on Phasmatodea continues. Our current project involves injecting Phasmatodea with a cocktail of pathogens in order to elicit an immune response. These immune-challenged Phasmatodea will be compared with healthy individuals, with RNA-seq hopefully identifying the antimicrobial peptides expressed in the former. Transcriptomics can also be used to solve similar unsolved questions in biology. One mystery dating back to 1950 involves the basking shark, *Cetorhinus maximus*, whose left ovary is non-functional. Only the right ovary produces eggs (Matthews, 1950). The question of why this phenomenon occurs has gone untouched since it was asked: perhaps a *de novo* transcriptome of the basking shark's female genitalia and RNA-seq between the left and right ovaries will answer this question. Another mystery dating back to 1949 is the glycogen body: a glycogen-rich structure in the spinal cord of birds that does not seem to be involved in energy storage, which is the normal function of glycogen (Doyle and Watterson, 1949). Its function, too, is unknown, despite birds being a considerably more thoroughly studied clade than the Phasmatodea. A transcriptomics assay will at least show what genes are expressed in this organ, which will hopefully reveal more than simply glycogen transport. I could not begin to hypothesize what transcriptomics studies on the shark ovary and glycogen body will find: the Phasmatodea have taught me that one can never guess what one will or will not find when mining such large data sets.

Despite considerable promise, transcriptomics data, especially *de novo* transcriptomes lacking a suitable reference genome, carry a risk that one will miss important features for a variety of reasons. These include totally novel genes that BLAST annotation cannot identify: indeed, a large number of genes from these transcriptomics studies often could not be matched with a known animal gene. Errors in the assembly are always a possibility: a combination of genome-guided and *de novo* assemblies is best (Lu *et al.*, 2013), but not always feasible when working with non-model organisms, and the computational burden and costs or availability of suitable software are not negligible issues. Multiple truncated transcripts may represent different parts of single genes, or different alleles will be identified as different transcripts, so certain assemblers may overestimate the true number of transcripts in a system. Lastly, if the sample tissue is lacking in any way (such as the 1KITE sample without abdomens), contaminated in any way (eg. with microbes, symbiotic or otherwise), or degraded in any way (as RNA degrades almost immediately after tissue death compared with DNA and is much harder to work with), then the transcriptome will be of reduced quality or helpfulness. Being mindful of these limitations is important, but the sheer discovery power of transcriptomes and immediate practical implications of the data compared with genomes, where expression is not guaranteed or quantifiable, makes the effort worth investing in.

Consider the state of Phasmatodea physiology knowledge less than a decade ago: the appendices' functions were as mysterious as they were a century prior, the remarkable horizontal pectinase gene transfer was unknown, and the multifunctional GH9 cellulases were thought to be impossible. Some of these discoveries could be made with targeted hypothesis testing, but others could never have been found unless someone looked, and the justification for looking would not have been to find them. Phasmatodea rewrote what we know about animal enzyme and insect fat metabolism, not to mention being fascinating in their own right, but we would have never made these discoveries without the wide net of transcriptome research. The initial search for Phasmatodea cellulases was a fishing expedition. The haul that came from it exceeded all expectations. While not all such expeditions will have the fortune to come across novel enzyme functions or lateral gene transfers, no discoveries can be made if nobody looks, so braving the seas of uncertainty to fish in the waters of organism RNA libraries is always worth the risk.

References

Bateman, A*et al.* (2004) The Pfam protein families database. *Nucleic Acids Research* 32(90001), D138–D141. DOI: 10.1093/nar/gkh121.

Bordas, L. (1896) Considérations générales sur l'appareil digestif des 'Phasmidae'. *Bulletin du Museum d'Histoire Naturelle (Paris)* 2, 378–379.

Boto, L. (2014) Horizontal gene transfer in the acquisition of novel traits by metazoans. *Proceedings of the Royal Society B: Biological Sciences* 281(1777), 20132450. DOI: 10.1098/rspb.2013.2450.

Bradler, S. (2015) Der Phasmatodea Tree of Life: Überraschendes und Ungeklärtes in der Stabschrecken-Evolution. *Entomologie Heute* 27, 1–23.

Bradler, S., Cliquennois, N. and Buckley, T.R. (2015) Single origin of the Mascarene stick insects: ancient radiation on sunken islands? *BMC Evolutionary Biology* 15(1), 196. DOI: 10.1186/s12862-015-0478-y.

Calderón-Cortés, N., Quesada, M., Watanabe, H., Cano-Camacho, H. and Oyama, K. (2012) Endogenous plant cell wall digestion: a key mechanism in insect evolution. *Annual Review of Ecology, Evolution, and Systematics* 43(1), 45–71. DOI: 10.1146/annurev-ecolsys-110411-160312.

Cameron, A.E. (1912) Structure of the alimentary canal of the stick-insect, *Bacillus rossii* Fabr.; with a note on the parthenogenesis of this species. *Proceedings of the Zoological Society of London* 82(1), 172–182. DOI: 10.1111/j.1469-7998.1912.tb07011.x.

Canavoso, L.E., Jouni, Z.E., Karnas, K.J., Pennington, J.E. and Wells, M.A. (2001) Fat metabolism in insects. *Annual Review of Nutrition* 21, 23–46.

Casadevall, A. and Fang, F.C. (2008) Descriptive Science. *Infection and Immunity* 76(9), 3835–3836. DOI: 10.1128/IAI.00743-08.

Davison, A. and Blaxter, M. (2005) Ancient origin of glycosyl hydrolase family 9 cellulase genes. *Molecular Biology and Evolution* 22(5), 1273–1284. DOI: 10.1093/molbev/msi107.

De Sinéty, R. (1901) Rechearches sur la biologie et l'anatomie des Phasmes. *La Cellule* 19, 117–278. DOI: 10.5962/bhl.part.17087.

Dow, J.A. (2009) Insights into the Malpighian tubule from functional genomics. *Journal of Experimental Biology* 212(3), 435–445. DOI: 10.1242/jeb.024224.

Doyle, W. and Watterson, R. (1949) The accumulation of glycogen in the 'glycogen body' of the nerve cord of the developing chick. *Journal of Morphology* 85(2), 391–403. DOI: 10.1002/jmor.1050850205.

Eklöf, J.M., Ruda, M.C. and Brumer, H. (2012) Distinguishing xyloglucanase activity in endo-beta (1–>4)glucanases. *Methods in Enzymology* 510, 97–120.

Fischer, R., Ostafe, R. and Twyman, R.M. (2013) Cellulases from insects. *Advances in Biochemical Engineering / Biotechnology* 136, 51–64.

Haas, B.J., Papanicolaou, A., Yassour, M., Grabherr, M., Blood, P.D. *et al.* (2013) De novo transcript sequence reconstruction from RNA-seq using the Trinity platform for reference generation and analysis. *Nature Protocols* 8(8), 1494–1512. DOI: 10.1038/nprot.2013.084.

Heymons, R. (1897) Ueber die Organisation und Entwickelung von *Bacillus rossii* Fabr. *Sitzungsberichte der Königlich Preussischen Akademie der Wissenschaften zu Berlin* 16, 363–373.

Hinton, H.E. (1981) *Biology of Insect Eggs*. Pergamon Press, Oxford, UK.

Huang, X. and Madan, A. (1999) CAP3: A DNA sequence assembly program. *Genome Research* 9(9), 868–877. DOI: 10.1101/gr.9.9.868.

Hughes, I., Binkley, J., Hurle, B., Green, E.D., Sidow, A. *et al.* (2008) Identification of the Otopetrin Domain, a conserved domain in vertebrate otopetrins and invertebrate otopetrin-like family members. *BMC Evolutionary Biology* 8(1), 41. DOI: 10.1186/1471-2148-8-41.

Kirsch, R., Gramzow, L., Theissen, G., Siegfried, B.D., ffrench-Constant, R.H. *et al.* (2014) Horizontal gene transfer and functional diversification of plant cell wall degrading polygalacturonases: key events in the evolution of herbivory in beetles. *Insect Biochemistry and Molecular Biology* 52C, 33–50. DOI: 10.1016/j.ibmb.2014.06.008.

Leng, N., Dawson, J.A., Thomson, J.A., Ruotti, V., Rissman, A.I. *et al.* (2013) EBSeq: An empirical Bayes hierarchical model for inference in RNA-seq experiments. *Bioinformatics* 29(8), 1035–1043. DOI: 10.1093/bioinformatics/btt087.

Letsch, H.O., Meusemann, K., Wipfler, B., Schutte, K., Beutel, R. *et al.* (2012) Insect phylogenomics: results, problems and the impact of matrix composition. *Proceedings of the Royal Society B: Biological Sciences* 279(1741), 3282–3290. DOI: 10.1098/rspb.2012.0744.

Li, B. and Dewey, C.N. (2011) RSEM: accurate transcript quantification from RNA-Seq data with or without a reference genome. *BMC Bioinformatics* 12(1), 323. DOI: 10.1186/1471-2105-12-323.

Lison, L. (1937) Études histophysiologiques sur les tubes de Malpighi des insectes. III. L'élimination des colorants basiques chez les orthoptères. *Cell and Tissue Research* 28, 179–209.

Lison, L. (1942) *Recherches sur l'histophysiologie comparée de l'excrétion chez les arthropodes.* Palais des académies, Brussels, Belgium.

Lu, B., Zeng, Z. and Shi, T. (2013) Comparative study of de novo assembly and genome-guided assembly strategies for transcriptome reconstruction based on RNA-Seq. *Science China Life Sciences* 56(2), 143–155. DOI: 10.1007/s11427-013-4442-z.

Mamidala, P., Wijeratne, A.J., Wijeratne, S., Kornacker, K., Sudhamalla, B. *et al.* (2012) RNA-Seq and molecular docking reveal multi-level pesticide resistance in the bed bug. *BMC Genomics* 13(1), 6. DOI: 10.1186/1471-2164-13-6.

Matthews, L.H. (1950) Reproduction in the basking shark, *Cetorhinus maximus* (Gunner). *Philosophical Transactions of the Royal Society of London B* 234, 247–316.

Monteiro, E.C., Tamaki, F.K., Terra, W.R. and Ribeiro, A.F. (2014) The digestive system of the 'stick bug' *Cladomorphus phyllinus* (Phasmida, Phasmatidae): a morphological, physiological and biochemical analysis. *Arthropod Structure and Development* 43(2), 123–134. DOI: 10.1016/j.asd.2013.11.005.

Pantel, J. (1919) Le calcium, forme de rêserve dans la femelle des Phasmides; ses formes d'élimination dans les deux sexes. *Comptes Rendus l'Académie des sciences. Paris* 168, 242–244.

Pauchet, Y., Freitak, D., Heidel-Fischer, H.M., Heckel, D.G. and Vogel, H. (2009) Immunity or digestion: glucanase activity in a glucan-binding protein family from Lepidoptera. *Journal of Biological Chemistry* 284, 2214–2224.

Pauchet, Y., Wilkinson, P., Chauhan, R. and Ffrench-Constant, R.H. (2010a) Diversity of beetle genes encoding novel plant cell wall degrading enzymes. *PLoS ONE* 5(12), e15635. DOI: 10.1371/journal.pone.0015635.

Pauchet, Y., Wilkinson, P., Vogel, H., Nelson, D.R., Reynolds, S.E. *et al.* (2010b) Pyrosequencing the *Manduca sexta* larval midgut transcriptome: messages for digestion, detoxification and defence. *Insect Molecular Biology* 19(1), 61–75. DOI: 10.1111/j.1365-2583.2009.00936.x.

Ramsay, J.A. (1954) Active transport of water by the Malpighian tubules of the stick insect, *Dixippus morosus* (Orthoptera, Phasmidae). *Journal of Experimental Biology* 31, 104–113.

Ramsay, J.A. (1955) The excretory system of the stick insect *Dixippus morosus* (Orthoptera, Phasmidae). *Journal of Experimental Biology* 32, 183–199.

Rowley, J.D. (1973) A new consistent chromosomal abnormality in chronic myelogenous leukaemia identified by quinacrine fluorescence and Giemsa staining. *Nature* 243(5405), 290–293. DOI: 10.1038/243290a0.

Savage, A.A. (1962) The development of the Malpighian tubules of *Carausius morosus* (Orthoptera). *Quarterly Journal of Microscopical Science* 103, 417–437.

Shelomi, M. (2017) *De novo* transcriptome analysis of the excretory tubules of *Carausius morosus* (Phasmatodea) and possible functions of the midgut 'appendices'. *PLoS ONE* 12(4), e0174984. DOI: 10.1371/journal.pone.0174984.

Shelomi, M. and Kimsey, L.S. (2014) Vital staining of the stick insect digestive system identifies appendices of the midgut as novel system of excretion. *Journal of Morphology* 275(6), 623–633. DOI: 10.1002/jmor.20243.

Shelomi, M., Lo, W.S., Kimsey, L.S. and Kuo, C.H. (2013) Analysis of the gut microbiota of walking sticks (Phasmatodea). *BMC Research Notes* 6(1), 368. DOI: 10.1186/1756-0500-6-368.

Shelomi, M., Jasper, W.C., Atallah, J., Kimsey, L.S. and Johnson, B.R. (2014a) Differential expression of endogenous plant cell wall degrading enzyme genes in the stick insect (Phasmatodea) midgut. *BMC Genomics* 15(1), 917. DOI: 10.1186/1471-2164-15-917.

Shelomi, M., Watanabe, H. and Arakawa, G. (2014b) Endogenous cellulase enzymes in the stick insect (Phasmatodea) gut. *Journal of Insect Physiology* 60, 25–30. DOI: 10.1016/j.jinsphys.2013.10.007.

Shelomi, M., Sitepu, I.R., Boundy-Mills, K.L. and Kimsey, L.S. (2015) Review of the gross anatomy and microbiology of the Phasmatodea digestive tract. *Journal of Orthoptera Research* 24(1), 29–40. DOI: 10.1665/034.024.0105.

Shelomi, M., Danchin, E.G.J., Heckel, D., Wipfler, B., Bradler, S. *et al.* (2016a) Horizontal gene transfer of pectinases from bacteria preceded the diversification of stick and leaf insects. *Scientific Reports* 6(1), 26388. DOI: 10.1038/srep26388.

Shelomi, M., Heckel, D.G. and Pauchet, Y. (2016b) Ancestral gene duplication enabled the evolution of multifunctional cellulases in stick insects (Phasmatodea). *Insect Biochemistry and Molecular Biology* 71, 1–11. DOI: 10.1016/j.ibmb.2016.02.003.

Shen, Z., Denton, M., Mutti, N., Pappan, K., Kanost, M.R. *et al.* (2003) Polygalacturonase from *Sitophilus oryzae*: possible horizontal transfer of a pectinase gene from fungi to weevils. *Journal of Insect Science* 3(24), 1–9. DOI: 10.1673/031.003.2401.

Soria-Carrasco, V., Gompert, Z., Comeault, A.A., Farkas, T.E., Parchman, T.L. *et al.* (2014) Stick insect genomes reveal natural selection's role in parallel speciation. *Science* 344(6185), 738–742. DOI: 10.1126/science.1252136.

Wang, G.X., Cho, K.W., Uhm, M., Hu, C.R., Li, S. *et al.* (2014) Otopetrin 1 protects mice from obesity-associated metabolic dysfunction through attenuating adipose tissue inflammation. *Diabetes* 63(4), 1340–1352. DOI: 10.2337/db13-1139.

10 A Practical Guide for Functional Transcriptomics: A Case Study in RNA Interference and qPCR to Understand the Explosive Chemistry of *Brachinus* Bombardier Beetles (Coleoptera: Carabidae)

Aman S. Gill,* Melanie Gee, and Kipling Will

Department of Environmental Science and Policy Management, University of California, Berkeley, USA

10.1 Introduction

Prospects for using transcriptome sequencing to uncover the genetic basis for phenotypes of interest in non-model insects have benefited tremendously from reduced sequencing costs and advances in computational and analytical approaches. Sequencing transcriptomes provides a biologically coherent way to reproducibly reduce the size of the genome, compared with more arbitrary approaches like using ultra-conserved elements or restriction-site associated loci. This allows researchers to sequence homologous loci across large numbers of samples. It is now within reach for even small laboratories to sequence transcriptomes with sufficient throughput to sample multiple species, populations, tissues, or experimental treatments, leading to applications in an increasing number of research fields.

The goal of this chapter is to communicate best practices in insect functional transcriptomics, defined as the use of transcriptomic data to identify and validate genes that express phenotypes of interest. Transcriptomics stands to benefit functional studies by providing a direct path to developing genome-wide toolboxes in almost any insect system. In a single experiment,

*Corresponding author: amango@gmail.com

for example, a researcher may obtain sequences, annotations, and gene expression values for all expressed genes (i.e. all genes transcribed into polyadenylated messenger RNA). When combined with robust experimental design, transcriptomics can then provide a basis for identifying and prioritizing candidate genes, and, critically, for validating the role of top candidates in functions (i.e. phenotypes) of interest. The initial analytical steps common to all transcriptome experiments, assembly and annotation, are well described, with many available tools for each. Here we describe a workflow for identifying candidate genes based on *de novo* transcriptome assemblies, functionally validating candidates using RNA interference (RNAi), and quantifying knockdown expression levels with quantitative PCR (qPCR). We emphasize 'universal' steps that could apply to any transcriptome study once an assembled, annotated transcriptome is obtained. These include design and delivery of short interfering RNA (siRNA) used for RNAi, identification of qPCR reference genes, and design of qPCR primers.

10.1.1 Case study system: *Brachinus* bombardier beetles

As a case study we present results from an investigation into the chemical defense system of the bombardier beetle *Brachinus elongatulus* Chaudoir, 1876 (Coleoptera: Carabidae). Carabidae, one of the largest families of animals, includes beetles with a single homologous defensive gland system that produces a strikingly diverse array of chemicals: 19 distinct classes representing >250 compounds, including hydrocarbons, formic acid, aromatic aldehydes, sulfur compounds, cyanides and quinones (Blum, 1981; Will *et al.*, 2000; Lorenz, 2005). Quinone-producing bombardiers stand out among this unparalleled chemical diversity with rapid-fire, precisely aimed explosive discharges that have been shown to deter even vertebrate predators (Eisner *et al.*, 1977; Dean *et al.*, 1990; Arndt *et al.*, 2015). The bombardier ability has almost certainly evolved multiple times in ground beetles, creating additional interest in understanding the genetic basis of bombardier defensive chemistry. As this chapter proceeds through the major steps in a functional transcriptome project, we first present general considerations for each step. We then describe specific methodological procedures as applied to our case study on a candidate gene for the concurrent oxidation of hydroquinone and decomposition of hydrogen peroxide: the powers of the bombardier beetle's explosive defenses. The emphasis here is to provide a practical example of a functional transcriptomics workflow, while the major results of this study will be reported in a related set of publications.

10.2 Generating Transcriptomic Data and Hypotheses – Transcriptome Assembly, Annotation, and Candidate Gene Identification

Any mass sequencing technology can be used for transcriptome sequencing, and any that reads sequences on a per molecule basis can be used for expression quantification based on sequence read counts (RNA-seq) (Torres *et al.*, 2007; Wilhelm and Landry, 2009). Likewise, several options are available for assembling quality-filtered reads into larger contigs – if sequencing depth is sufficient, contigs should resemble full-length transcripts. Assembly may make use of a related reference genome as a guide or reads may be assembled into *de novo* contigs (Simpson *et al.*, 2009; Robertson *et al.*, 2010; Surget-Groba and Montoya-Burgos, 2010; Grabherr *et al.*, 2011). The completeness of a reference-enabled or *de novo* assembly can be assessed using BUSCO scores (Benchmarking Universal Single-Copy Orthologs) (Simão *et al.*, 2015), and by evaluating the number, length, and similarity of transcripts relative to related organisms (O'Neil and Emrich, 2013). Several software packages are available to annotate transcripts based on sequence or motif similarity to one or more curated databases of orthologous proteins from representative taxa (Conesa *et al.*, 2005).

10.2.1 *B. elongatulus* assembly and annotation

Sequencing, assembly, and annotation of the *B. elongatulus* transcriptome are described separately (Renner *et al.*, in preparation). In brief, trimmed paired-end reads were assembled by Trinity 2.3.2 into *de novo* contigs (Haas *et al.*, 2013). Transcriptome quality and completeness were assessed using N50 contig lengths and BUSCO scores, using the Arthropoda *odb9* dataset (Simão *et al.*, 2015). Assembled transcripts were annotated using the Trinotate pipeline (v.3.0.2) by comparison with Gene Ontology (GO), Pfam, NCBI, and other databases.

10.2.2 Candidate gene prioritization

There are no universal approaches to identifying and prioritizing candidate genes. Candidate gene identification should take into account the biological system, particularly the availability of existing genomic tools, and the potential presence of similar functional studies in related taxa (Börnigen *et al.*, 2012). Approaches will also differ according to the function of interest – chemical and behavioral phenotypes, to take just two examples, would benefit from integration of very different assays alongside gene expression and annotation data. In comparative studies the axis of comparison will determine how to make use of expression data. If multiple species are involved, then comparing whole-organism wildtype expression levels may be sufficient. If the research question pertains more to genes in a certain tissue type or treatment for a single species, then differential expression across tissues or experimental treatments will be more appropriate.

10.2.3 Candidate genes for explosive hydroquinone oxidation

To identify candidate genes involved in the biosynthetic pathways underlying bombardier chemical defense, we integrated three types of data: gene expression in defensive gland tissues, orthology to proteins with known or suspected roles in benzoquinone synthesis, and results from incorporation of labeled chemical precursors. Gene expression data were derived from two distinct experiments. First, we examined the fold change of transcripts in two tissues integral to the defensive gland system in bombardiers: secretory cells (SCs) and reaction chambers (RCs). Methods and results are described by Renner *et al.* (in preparation), but briefly RSEM and edgeR were used to identify genes that were highly differentially expressed (fold change >2) in SCs and/or RCs compared with non-defensive gland tissues (Robinson *et al.*, 2010; Li and Dewey, 2011). In the second experiment, beetles were caused to exhaust their defensive reserves, ostensibly initiating defensive compound biosynthesis, and then SCs and RCs from replicated groups were sampled after 0.5, 1, 6, 12, 24 and 48h (hereafter referred to as the time series experiment). RSEM (Li and Dewey, 2011) was used to estimate transcript abundance, and the limma R/Bioconductor package was used to identify differentially expressed transcripts (Ritchie *et al.*, 2015). Co-expression relationships (i.e. transcript clusters with significantly positively and negatively correlated expression across time points) were also examined using CoExpNetViz (Tzfadia *et al.*, 2016). Transcripts with comparatively high expression in defensive gland tissues were noted and cross-referenced with results from orthology identification and chemical incorporations, as described below.

Proteins with known or suspected roles in benzoquinone biosynthesis were identified from multiple sources. Proteins from confused flour beetle *Tribolium castaneum* (Herbst, 1797) (Coleoptera: Tenebrionidae), which, like bombardier beetles, produces benzoquinones, were identified based on a study of defensive gland proteins (Li *et al.*, 2013) and the iBeetle database of RNAi phenotypes (Schmitt-Engel *et al.*, 2015). Other proteins were identified based on involvement in benzoquinone pathways in other organisms (Meinwald *et al.*, 1966; Rocha *et al.*, 2013). Once a database of putative defensive biosynthesis genes was compiled, profile hidden Markov models (HMMs) were obtained from the Pfam database (pfam.xfam.org) or generated anew using the HMMer software (hmmer.org). Compared with BLAST-like searches based on nucleotide or amino acid sequences, HMM searches are more effective at detecting orthology despite sequence divergence by modeling conserved functional domains. *B. elongatulus* open reading frames (ORFs) with high similarity to one or more HMMs were identified, resulting in a list of pre-candidate genes potentially encoding functions related to defensive compound biosynthesis.

The identity of the benzoquinone compounds secreted by *Brachinus* bombardier beetles have been known for decades after first being identified by mass spectrometry (Schildknecht *et al.*, 1968; Eisner *et al.*, 1977). The benzoquinone precursors have not been identified, nor is it clear to what degree various benzoquinone pathways share a common pathway. In order to identify chemical precursors of the 1,4-benzoquinone and toluquinone produced by *B. elongatulus*, we fed live beetles labeled doses of several hydroquinone compounds, as well as m-cresol and several other potential quinone precursors. As described fully in Attygalle *et al.* (in preparation), we established that m-cresol is converted to toluquinone, and hydroquinone is converted to 1,4-benzoquinone.

Our candidate gene prioritization process resulted in approximately two dozen candidates putatively involved in steps in either benzoquinone pathway (Renner *et al.*, in preparation). For the remainder of the chapter we focus on a set of transcripts encoding a cyclo-oxygenase-2 protein, also known as prostaglandin H synthase (*pghs*). These proteins contain a peroxidase site in addition to an active site capable of oxidation, consistent with a role in the exothermic decomposition of hydrogen peroxide and oxidation of hydroquinone into benzoquinone. This reaction takes place in the reaction chambers, producing the explosive blast the beetle is able to direct using features of its external morphology (Eisner and Aneshansley, 1982; Dean *et al.*, 1990; Arndt *et al.*, 2015).

10.3 Functional Validation for Non-Model Insect Transcriptomics

10.3.1 RNA interference

Several methods are available for functional validation, including the widely used CRISPR-cas9 system. The use of CRISPR is limited, however, when blocking gene function is required *in vivo*, e.g. in the absence of reproducing laboratory colonies. RNA interference (RNAi) can be used in such cases. Here we provide a practical guide to the use of RNAi based on *de novo* transcriptomic data.

10.3.1.1 Challenges of RNAi in insects

RNAi is a technique for depleting the messenger RNA (mRNA) transcript of a target gene, allowing observation of any resulting loss-of-function phenotypes. The mRNA transcript is targeted for degradation using double-stranded RNA (dsRNA) with a strand complementary to a region of the mRNA. The enzyme Dicer cleaves the mRNA into short interfering RNAs (siRNAs), which become integrated into the RNA-induced silencing complex (RISC) (Yu *et al.*, 2012). RISC binds to and cuts a homolog mRNA, preventing it from being translated. RNAi has been used to study insect embryonic development, postembryonic development, reproduction, behavior, biosynthetic pathways, and resistance to chemical and biological agents (Bellés, 2010).

Developing an RNAi project targeting candidate genes in a system lacking genomic resources requires addressing at least four broad practical challenges: designing RNAi constructs, delivering constructs into the organism, quantifying expression of genes targeted for knockdown, and assaying knockdown and control phenotypes.

10.3.1.2 Designing RNAi constructs

The first consideration in planning an RNAi treatment is to determine the nature of the RNA duplex that will be administered for incorporation into the endogenous RNAi pathway. A common approach is to use long double-stranded RNA molecules reverse transcribed from cDNA that is enriched for the target gene. The long dsRNAs are cleaved by Dicer enzymes into short 19bp duplexes, which are incorporated as guide molecules into RISCs that degrade the target RNA (Meister and Tuschl, 2004). An alternative, increasingly popular approach is to use synthetic RNA duplexes (siRNAs) as RNAi reagents, effectively skipping Dicer processing to directly provide guide molecules for RISCs (Kim *et al.*, 2005; Peek and Behlke, 2007). This is the approach we discuss here.

Several considerations influence siRNA design, beginning with the target transcript nucleotide sequence. In most cases detailed gene models will not be available for organisms lacking annotated genomes, so target transcripts must be identified based on transcriptome assemblies. Candidate genes identified in transcriptome experiments are often represented by clusters of isoforms, leading to multiple potential target sites for a given gene. Gene isoforms (i.e. transcripts sharing a subset of putative exons) should be clustered and examined to identify the primary target exon(s), which may be represented in multiple transcripts. Once a target exon is identified, effective siRNA sites should be chosen for thermodynamic stability, absence of homopolymeric runs or long G/C regions, and appropriate secondary structure (Kim *et al.*, 2005; Peek and Behlke, 2007). The easiest way to ensure these criteria are met is to generate siRNAs using a design tool, e.g. the Custom Dicer-Substrate siRNA tool made available by Integrated DNA Technologies. This and similar tools will report potential RNAi constructs ranked according to the above criteria. Constructs should be screened to ensure lack of non-target matches across the transcriptome. It is advisable to select multiple constructs per target to maximize the knockdown effect and to replicate results.

Controls are a critical consideration in RNAi treatments. The effects of siRNA load can be controlled using a 'scrambled' siRNA molecule with low similarity to all potential target transcripts. A negative control for delivery vehicle (siRNA solvent or medium) can be achieved by adding a null treatment (e.g. injections or feed lacking siRNAs) alongside the RNAi treatment(s). If available, positive controls (RNAi targets with known phenotypes) enhance the reliability and replicability of RNAi experiments. In general, positive controls require previous attempts at RNAi in the same or closely related organismal system.

10.3.1.3 siRNA design for *B. elongatulus pghs* gene

For the *pghs* gene, we identified five candidate transcript isoforms. To identify target regions for RNAi constructs, isoforms were clustered based on sequence similarity using the usearch cluster_fast algorithm (Edgar, 2010), resulting in two distinct clusters, *pghs1* and *pghs2*. Target regions were then submitted to the online IDT Custom Dicer-Substrate siRNA tool (www.idtdna.com/site/order/designtool/index/DSIRNA_CUSTOM) to design RNAi constructs (i.e. dicer-substrate siRNAs) such that each of the five isoforms would have unique matches to two constructs. This resulted in five siRNAs targeting *pghs1* (which included three isoforms) and two siRNAs for *pghs2*. As a negative RNAi control, one of the *pghs* siRNAs was scrambled using a GenScript tool (www.genscript.com/tools/create-scrambled-sequence), resulting in a scrambled siRNA with the same length, base composition, and complexity as the input siRNA. Each siRNA sequence was screened to ensure lack of high identity matches to other *B. elongatulus* transcripts with usearch search_oligodb, minimizing the risk of off-target effects. The six RNAi constructs were synthesized by IDT.

10.3.1.4 RNAi delivery

Three main challenges of inducing RNAi in non-model insects are delivery of the RNAi molecule into the insect, uptake of the RNAi molecule in the insect, and differing degrees of sensitivity to RNAi depending on the species, tissue, or even gene of interest. Common methods of RNAi delivery are microinjections, feeding, and soaking. Microinjection of the RNAi molecule (usually dsRNA) with needles allows a researcher to insert an exact, known amount of dsRNA into the tissue of choice, although factors such as insertion location, volume, and needle choice must be first optimized for successful delivery (Yu *et al.*, 2012). Feeding, also known as oral delivery, involves synthesizing dsRNA *in vitro* or expressing dsRNA in bacteria, and then adding the dsRNA molecule to the insect's diet. Advantages of feeding are that it is less invasive and easier to perform than microinjections; disadvantages of feeding are that it limits RNAi to only gut cells in species where systemic RNAi cannot be achieved (Scott *et al.*, 2013). Soaking involves literally soaking the insect or cells in a solution of dsRNA, and is a common method of RNAi delivery in insect cell lines rather than the organism itself (Yu *et al.*, 2012).

Uptake of the RNAi molecule in the insect refers to the efficacy with which cells uptake the delivered dsRNA, which can differ by species and tissue (Scott *et al.*, 2013). Once the dsRNA is in the cell, the target mRNA may be too transient for efficient RNAi or may be protected against enzymes that degrade RNA (Bellés, 2010).

10.3.2 Transcriptome-Based qPCR

Accurately measuring gene expression is a crucial step in an RNAi experiment to ensure that transcription has indeed been knocked down. Expression can be quantified using any available method – including qPCR, microarray, and RNA-seq. qPCR, also known as quantitative real-time PCR (qRT-PCR), is most often used due to its ease and low cost in targeting expression of just one or several genes, a typical scale for RNAi treatments. The key steps in developing accurate qPCR-based gene expression assays are establishing reference genes that are stably expressed across the conditions or treatments of interest in a given study, designing and validating qPCR primers for both target candidate genes and reference genes, and generating and analyzing raw qPCR data in order to quantify the expression of target genes in RNAi versus control treatments, relative to reference gene expression. To ensure reliable, replicable qPCR results, investigators should verify that assay procedures and reporting conform to the Minimum Information for Publication of Quantitative Real-Time PCR Experiments (MIQE) guidelines (Bustin *et al.*, 2009, 2010).

10.3.2.1 RNA isolation

Obtaining high-quality RNA is a critical step both for obtaining high-quality transcriptome assemblies and for generating reliable samples used for downstream functional validation using qRT-PCR. High-quality RNA should be of sufficient quantity as measured by spectrometry (Nanodrop or Qubit readings), integrity as indicated by Bioanalyzer traces, and purity as determined by the absence of contaminating genomic DNA.

Integrity , or intactness can typically be quantified with the RNA integrity number (RIN), which is a ratio of the 28S rRNA peak and the 18S peak, with RIN values greater than 2 indicating RNA of good integrity (Schroeder *et al.*, 2006). RIN scores are often derived from Bioanalyzer analysis of the distribution of fragment sizes. However, in beetles and other insects, the 28S rRNA sequence contains a natural scission, also called the 'hidden break' or 'gap deletion', that denatures upon heating (McCarthy *et al.*, 2015). The 28S fragments migrate with the 18S rRNA sequence, and a Bioanalyzer profile displays only one peak rather than the usual two (Winnebeck *et al.*, 2010). Since the RIN is a ratio of the two 28S and 18S peaks, it is not an appropriate measure of RNA integrity in these cases. Instead the size distribution should be manually examined to confirm the lack of short, degraded molecules alongside the presence of a major peak or two adjacent peaks representing the 18S and dual 28S peaks. All RNA samples used for downstream analysis should be confirmed to have similar size distribution.

10.3.2.2 Selecting and validating reference genes for qPCR

Experimental and biological variation are inevitable complications in qPCR experiments. This necessitates normalization strategies using reference genes with relatively stable expression across individuals as well as the treatments or conditions under study. Because reference genes will themselves exhibit innate variability (among individuals, tissues, etc.), investigators should use multiple reference genes in qPCR assays (Guenin *et al.*, 2009; Chervoneva *et al.*, 2010). Reference genes should be determined empirically for each study unless the experimental set-up is highly similar to existing qPCR assays in the same organismal system – the previously common practice of using one of a small set of putative 'housekeeping' genes has been shown to lead to significant error, as many standard housekeeping genes have modified expression across treatments or stochastic variation among samples (Thellin *et al.*, 1999; Schmittgen and Zakrajsek, 2000). Several methods have been introduced to rank potential reference genes among a pool of candidates based on expression stability across samples (Vandesompele *et al.*, 2002) that may also take into account variation across treatments (Andersen *et al.*, 2004) and possible correlation between reference genes (Chervoneva *et al.*, 2010).

We used the geNorm algorithm (Vandesompele *et al.*, 2002) to identify reference genes based on RNA-seq data from *Brachinus* reaction chamber tissues derived from 18 individuals sampled in triplicate from each of six time points spanning defensive chemical biosynthesis (described above). Genes with mean TMM-normalized abundance below ten were excluded from consideration. The coefficient of variation (CV) was calculated for the remaining transcripts as CV = SD/mean. Transcripts with CV <0.2 and fold change <0.4 were sorted based on mean abundance, and the five genes with the highest mean TMM values were selected for qPCR primer design and validation. In addition to these transcriptome-based candidate reference genes, we investigated reference genes used in published qPCR studies in other beetles: *ActB*, *CypA*, *Ef1a*, *RpL13a*, *RpS18*, *Sps3*, *TAF5*, *UBE3A*, *gapdh* (Lord *et al.*, 2010; Shi *et al.*, 2013; Wang *et al.*, 2014; Pan *et al.*, 2015; Tan *et al.*, 2015). For each of these genes, *Brachinus* orthologs were identified using tblastx searches against the full set of transcriptome sequences. Primers for all candidate reference genes were designed in primer3 (Untergasser *et al.*, 2012) using default qPCR settings and also specifying a minimum primer length of 80, maximum length of 150, and optimal length of 100. Potential primers were screened against the entire *B. elongatulus* transcriptome to ensure target specificity and lack of off-target hits, using usearch search_oligodb. These primers were then used for qPCR on a common set of samples, and the resulting quantification cycle (C_q) values were run through the geNorm algorithm (Vandesompele *et al.*, 2002) as implemented in the R package NormqPCR (Perkins *et al.*, 2012).

10.3.2.3 qPCR assay optimization and validation

Once candidate genes and reference genes are established and primers are designed, PCR amplification efficiency and other assay performance characteristics should be determined. Target and reference genes must be amplified with similar efficiency in order to accurately estimate concentration differences between samples for the $\Delta\Delta C_q$ method, described below. Efficiency is established based on the log-linear portion of a calibration curve derived from a dilution series of a common sample. A 100% efficiency indicates that the amplicon concentration doubles with each cycle, but values between 90% and 110% are often considered acceptable. Dilution series can also be used to establish the dynamic range of the assay, allowing investigators to dilute template samples and still ensure precise quantification. Primer specificity should also be established either by sequencing amplicons or by examining melting profiles based on a qPCR run.

Candidate qPCR primers were tested for efficiency by running qPCR reactions on serial dilutions of a sample and observing if amplification was proportional to the dilutions. For each primer pair and dilution, qPCR reactions were done in triplicate on a 96-well plate. A sample already diluted 1:5 was serially diluted fivefold another five times, resulting in the following dilutions: 1:5, 1:25, 1:125, 1:625, 1:3125, 1:15625. A non-template control (NTC) was also included in triplicate for each primer pair. qPCR reactions were assembled and run as described below for RNAi-treated and control samples. For a given primer pair, triplicate fluorescence values of each dilution were averaged and graphed linearly, and efficiencies were calculated to ensure that values were in the range of 90–100%. Melt curves were also examined to confirm that fluorescence values were based on a single amplification of a single fragment.

10.3.2.4 Generating and analyzing qPCR data

qPCR reactions are assembled in much the same way as standard PCR reactions, with the exception that a fluorescent reporter that binds to DNA is included, causing fluorescence to increase in direct proportion to the log-linear increase in target template concentration. The point at which the fluorescence becomes measurable above the baseline level is the quantification cycle (C_q), also known as quantification threshold (C_t). Gene expression can be quantified as the relative fold change based directly on C_q values. The mean C_q (of triplicates for each sample) of a target gene minus that of the top ranked reference gene (see above) gives the ΔC_q. The ΔC_q of a target gene from experimental samples can be subtracted from that of control samples to calculate the $\Delta\Delta C_q$, and fold change is calculated as $2^{-\Delta\Delta Cq}$. Fold change values below 1 indicate reduced expression, providing a direct measurement of the presence and degree of expression knockdown by RNAi.

For each primer pair and sample, qPCR reactions were done in triplicate on a 96-well plate. Samples were diluted 1:5. The master mix consisted of (per sample): 2 µl 10x PCR Rxn Buffer -Mg^{2+} (Invitrogen), 0.8 µl 50 mM $MgCl_2$ (Invitrogen), 0.4 10 µM dNTPs (Invitrogen), 0.4 µl Rox Reference Dye (Invitrogen), 0.2 µl SYBR Green I (Thermo Fisher), 0.1 µl 5 U/µl *Taq* DNA Polymerase (Invitrogen), and 13.3 µl DNase/RNase-free water. Each qPCR reaction consisted of 18 µl master mix per well and 2 µl sample per well (total reaction volume: 20 µl). The qPCR reactions were run in an Applied Biosystems StepOnePlus™ Real-Time PCR System (Applied Biosystems). The program settings were as follows. Instrument: StepOnePlus™ Instrument (96 wells); type of experiment: Quantitation – Relative Standard Curve; reagents to detect target sequence: SYBR Green Reagents; ramp speed: standard (~2 h to complete a run). The thermocycler parameters were as follows: holding stage 95°C for 10 min; cycle stage 40 cycles of 95°C for 15 s, 60°C for 30 s, 72°C for 30 s; melt curve stage 95°C for 15 s, 60°C for 1 min.

10.3.3 Phenotypic Assays

Validating candidate genes ultimately requires establishing a causal relationship between gene expression and change in one or more phenotypes. Developing phenotypic assays that allow quantitative or qualitative diagnosis of discrete phenotypes is a critical step in functional transcriptomics. Assays will vary widely with different organismal systems and research questions. If possible, multiple assays should be designed to provide independent confirmation of phenotypic changes.

We used gas chromatography–mass spectrometry (GC–MS) and enzyme kinetics to assess chemical phenotypes in replicated groups of *pghs* RNAi knockdowns and control *B. elongatulus* individuals. Details of these assays are reported separately (Gill *et al.*, unpublished); briefly, for each individual beetle in the RNAi trial we sampled chemical blasts at four time points during the 2-week feeding trial. The liquid portion of each blast was collected in a tube and washed with a GC–MS-compatible solvent, in this case ethyl acetate with 8 mM trifluorotoluene as an internal standard. Once the RNAi feed trial concluded, beetles were briefly anesthetized in a freezer and reaction chambers were dissected and placed into a solution of RNAlater to allow dual use for enzyme kinetics and RNA isolation for qPCR as described above. Standard GC–MS curves were generated based on serial dilutions of commercially synthesized compounds present in reaction chambers: hydroquinone, methyl-hydroquinone, 1,4-benzoquinone, and toluquinone (methyl-1,4-benzoquinone). GC–MS spectra were then obtained for each sample, allowing relative quantification of all four compounds. For the kinetics, *in vitro* reactants were established alongside a protocol to isolate reaction chamber enzymes from *B. elongatulus* individuals. Enzyme solution was added to reactants in a replicable manner using a stopped flow device, and the absorbance spectra of the resulting reaction was recorded by a UV spectrometer, allowing calculation of the enzyme's catalytic rate.

10.4 Conclusion

Advances in sequencing technologies and analytical methods have made transcriptomic approaches in insects more accessible than ever, paving the way for novel discoveries in the genetic basis and evolutionary history of functional traits in under-explored branches of the tree of life. Presently, functions encoded by one or several genes will be the most amenable to transcriptome-based functional genetic approaches. As developing sequencing technologies allow sequencing of increasingly large read lengths, transcriptome and genome assemblies will improve, allowing for greater understanding of functions encoded and regulated by more complex genetic architectures.

Acknowledgements

This work was funded by Natural Science Foundation (NSF) Grant 1556957 to KW. Discussions with the Carabid-Q team were indispensable: Tanya Renner, Wendy Moore, Athula Attygalle, Reilly McManus, Sihang Xu, Sean Perez, Anita Silver, Betahny Xu, Shreeya Garg, and Jerry Yin.

References

Andersen, C.L., Jensen, J.L. and Ørntoft, T.F. (2004) Normalization of real-time quantitative reverse transcription-PCR data: a model-based variance estimation approach to identify genes suited for normalization, applied to bladder and colon cancer data sets. *Cancer Research* 64(15), 5245–5250. DOI: 10.1158/0008-5472.CAN-04-0496.

Arndt, E.M., Moore, W., Lee, W.-K. and Ortiz, C. (2015) Mechanistic origins of bombardier beetle (Brachinini) explosion-induced defensive spray pulsation. *Science* 348(6234), 563–567. DOI: 10.1126/science.1261166.

Bellés, X. (2010) Beyond Drosophila: RNAi in vivo and functional genomics in insects. *Annual Review of Entomology* 55(1), 111–128. DOI: 10.1146/annurev-ento-112408-085301.

Bustin, S.A., Beaulieu, J.-F., Huggett, J., Jaggi, R., Kibenge, F.S.B. *et al.* (2010) MIQE précis: Practical implementation of minimum standard guidelines for fluorescence-based quantitative real-time PCR experiments. *BMC Molecular Biology* 11(1), 74. DOI: 10.1186/1471-2199-11-74.

Bustin, S.A., Benes, V., Garson, J.A., Hellemans, J., Huggett, J. *et al.* (2009) The MIQE Guidelines: Minimum Information for Publication of Quantitative Real-Time PCR Experiments. *Clinical Chemistry* 55(4), 611–622. DOI: 10.1373/clinchem.2008.112797.

Blum, M.S. (1981) Arthropod defensive compounds. In: Blum, M.S. (ed.) *Chemical Defenses of Arthropods*. Elsevier Inc, Amsterdam, Netherlands, pp. 496–511.

Bornigen, D., Tranchevent, L.-C., Bonachela-Capdevila, F., Devriendt, K., De Moor, B. *et al.* (2012) An unbiased evaluation of gene prioritization tools. *Bioinformatics* 28(23), 3081–3088. DOI: 10.1093/bioinformatics/bts581.

Chervoneva, I., Li, Y., Schulz, S., Croker, S., Wilson, C. *et al.* (2010) Selection of optimal reference genes for normalization in quantitative RT-PCR. *BMC Bioinformatics* 11(1), 253. DOI: 10.1186/1471-2105-11-253.

Conesa, A., Götz, S., García-Gómez, J.M., Terol, J., Talón, M. *et al.* (2005) Blast2GO: a universal tool for annotation, visualization and analysis in functional genomics research. *Bioinformatics* 21(18), 3674–3676. DOI: 10.1093/bioinformatics/bti610.

Dean, J., Aneshansley, D.J., Edgerton, H.E. and Eisner, T. (1990) Defensive spray of the bombardier beetle - a biological pulse jet. *Science* 248(4960), 1219–1221. DOI: 10.1126/science.2349480.

Edgar, R.C. (2010) Search and clustering orders of magnitude faster than BLAST. *Bioinformatics* 26(19), 2460–2461. DOI: 10.1093/bioinformatics/btq461.

Eisner, T. and Aneshansley, D.J. (1982) Spray aiming in bombardier beetles – jet deflection by the coanda effect. *Science* 215(4528), 83–85. DOI: 10.1126/science.215.4528.83.

Eisner, T., Jones, T.H., Aneshansley, D.J., Tschinkel, W.R., Silberglied, R.E. *et al.* (1977) Chemistry of defensive secretions of bombardier beetles (Brachinini, Metriini, Ozaenini, Paussini). *Journal of Insect Physiology* 23(11–12), 1383–1386. DOI: 10.1016/0022-1910(77)90162-7.

Grabherr, M.G., Haas, B.J., Yassour, M., Levin, J.Z., Thompson, D.A. *et al.* (2011) Full-length transcriptome assembly from RNA-Seq data without a reference genome. *Nature Biotechnology* 29(7), 644–652. DOI: 10.1038/nbt.1883.

Guenin, S., Mauriat, M., Pelloux, J., Van Wuytswinkel, O., Bellini, C. *et al.* (2009) Normalization of qRT-PCR data: the necessity of adopting a systematic, experimental conditions-specific, validation of references. *Journal of Experimental Botany* 60(2), 487–493. DOI: 10.1093/jxb/ern305.

Haas, B.J., Papanicolaou, A., Yassour, M., Grabherr, M., Blood, P.D. *et al.* (2013) De novo transcript sequence reconstruction from RNA-seq using the Trinity platform for reference generation and analysis. *Nature Protocols* 8(8), 1494–1512. DOI: 10.1038/nprot.2013.084.

Kim, D.-H., Behlke, M.A., Rose, S.D., Chang, M.-S., Choi, S. *et al.* (2005) Synthetic dsRNA Dicer substrates enhance RNAi potency and efficacy. *Nature Biotechnology* 23(2), 222–226. DOI: 10.1038/nbt1051.

Li, B. and Dewey, C. (2011) RSEM: accurate transcript quantification from RNA-Seq data with or without a reference genome. *BMC Bioinformatics* 12(1), 323. DOI: 10.1186/1471-2105-12-323.

Li, J., Lehmann, S., Weißbecker, B., Ojeda Naharros, I., Schütz, S. *et al.* (2013) Odoriferous defensive stink gland transcriptome to identify novel genes necessary for quinone synthesis in the red flour beetle, *Tribolium castaneum*. *PLoS Genetics* 9(7), e1003596. DOI: 10.1371/journal.pgen.1003596.

Lord, J.C., Hartzer, K., Toutges, M. and Oppert, B. (2010) Evaluation of quantitative PCR reference genes for gene expression studies in *Tribolium castaneum* after fungal challenge. *Journal of Microbiological Methods* 80(2), 219–221. DOI: 10.1016/j.mimet.2009.12.007.

Lorenz, W. (2005) *Systematic List of Extant Ground Beetles of the World*, 2nd edn. Tutzing, Germany.

McCarthy, S.D., Dugon, M.M. and Power, A.M. (2015) 'Degraded' RNA profiles in Arthropoda and beyond. *PeerJ* 3(1), e1436. DOI: 10.7717/peerj.1436.

Meinwald, J., Koch, K.F., Rogers, J.E. Jr. and Eisner, T. (1966) Biosynthesis of arthropod secretions. III. Synthesis of simple p-benzoquinones in a beetle (*Eleodes longicollis*). *Journal of the American Chemical Society* 88(7), 1590–1592. DOI: 10.1021/ja00959a074.

Meister, G. and Tuschl, T. (2004) Mechanisms of gene silencing by double-stranded RNA. *Nature* 431(7006), 343–349. DOI: 10.1038/nature02873.

O'Neil, S.T. and Emrich, S.J. (2013) Assessing De Novo transcriptome assembly metrics for consistency and utility. *BMC Genomics* 14(1), 465. DOI: 10.1186/1471-2164-14-465.

Pan, H., Yang, X., Siegfried, B.D. and Zhou, X. (2015) A comprehensive selection of reference genes for RT-qPCR analysis in a predatory lady beetle, *Hippodamia convergens* (Coleoptera: Coccinellidae). *PLoS ONE* 10(4), e0125868. DOI: 10.1371/journal.pone.0125868.

Peek, A.S. and Behlke, M.A. (2007) Design of active small interfering RNAs. *Current Opinion in Molecular Therapeutics* 9(2), 110–118.

Perkins, J.R., Dawes, J.M., McMahon, S.B., Bennett, D.L.H., Orengo, C. *et al.* (2012) ReadqPCR and NormqPCR: R packages for the reading, quality checking and normalisation of RT-qPCR quantification cycle (Cq) data. *BMC Genomics* 13(1), 296. DOI: 10.1186/1471-2164-13-296.

Ritchie, M.E., Phipson, B., Wu, D., Hu, Y., Law, C.W. *et al.* (2015) Limma powers differential expression analyses for RNA-sequencing and microarray studies. *Nucleic Acids Research* 43(7), e47. DOI: 10.1093/nar/gkv007.

Robertson, G., Schein, J., Chiu, R., Corbett, R., Field, M. *et al.* (2010) De novo assembly and analysis of RNA-seq data. *Nature Methods* 7(11), 909–912. DOI: 10.1038/nmeth.1517.

Robinson, M.D., McCarthy, D.J. and Smyth, G.K. (2010) edgeR: a Bioconductor package for differential expression analysis of digital gene expression data. *Bioinformatics* 26(1), 139–140. DOI: 10.1093/bioinformatics/btp616.

Rocha, D., Wouters, F., Zampieri, D., Brocksom, T., Machado, G. *et al.* (2013) Harvestman phenols and benzoquinones: characterisation and biosynthetic pathway. *Molecules* 18(9), 11429–11451. DOI: 10.3390/molecules180911429.

Schildknecht, H., Maschwitz, E. and Maschwitz, U. (1968) [XXXIV. On arthropod protective substances. Explosion chemistry of the bombardier beetle (Coleoptera, Carabidae). 3. Isolation and characterization of the explosion catalysts]. *Zeitschrift fur Naturforschung.* 23(9), 1213–1218.

Schmitt-Engel, C., Schultheis, D., Schwirz, J., Ströhlein, N., Troelenberg, N. *et al.* (2015) The iBeetle large-scale RNAi screen reveals gene functions for insect development and physiology. *Nature Communications* 6(1), 1–10. DOI: 10.1038/ncomms8822.

Schmittgen, T.D. and Zakrajsek, B.A. (2000) Effect of experimental treatment on housekeeping gene expression: validation by real-time, quantitative RT-PCR. *Journal of Biochemical and Biophysical Methods* 46(1-2), 69–81. DOI: 10.1016/S0165-022X(00)00129-9.

Schroeder, A., Mueller, O., Stocker, S., Salowsky, R., Leiber, M. *et al.* (2006) The RIN: an RNA integrity number for assigning integrity values to RNA measurements. *BMC Molecular Biology* 7(1), 3. DOI: 10.1186/1471-2199-7-3.

Scott, J.G., Michel, K., Bartholomay, L.C., Siegfried, B.D., Hunter, W.B. *et al.* (2013) Towards the elements of successful insect RNAi. *Journal of Insect Physiology* 59(12), 1212–1221. DOI: 10.1016/j.jinsphys.2013.08.014.

Simpson, J.T., Wong, K., Jackman, S.D., Schein, J.E., Jones, S.J.M. *et al.* (2009) ABySS: a parallel assembler for short read sequence data. *Genome Research* 19(6), 1117–1123. DOI: 10.1101/gr.089532.108.

Shi, X.-Q., Guo, W.-C., Wan, P.-J., Zhou, L.-T., Ren, X.-L. *et al.* (2013) Validation of reference genes for expression analysis by quantitative real-time PCR in Leptinotarsa decemlineata (Say). *BMC Research Notes* 6(1), 1. DOI: 10.1186/1756-0500-6-93.

Simão, F.A., Waterhouse, R.M., Ioannidis, P., Kriventseva, E.V. and Zdobnov, E.M. (2015) BUSCO: assessing genome assembly and annotation completeness with single-copy orthologs. *Bioinformatics* 31(19), 3210–3212. DOI: 10.1093/bioinformatics/btv351.

Surget-Groba, Y. and Montoya-Burgos, J.I. (2010) Optimization of de novo transcriptome assembly from next-generation sequencing data. *Genome Research* 20(10), 1432–1440. DOI: 10.1101/gr.103846.109.

Tan, Q.-Q., Zhu, L., Li, Y., Liu, W., Ma, W.-H. *et al.* (2015) A de novo transcriptome and valid reference genes for quantitative real-time PCR in *Colaphellus bowringi*. *PLoS ONE* 10(2), e0118693. DOI: 10.1371/journal.pone.0118693.

Thellin, O., Zorzi, W., Lakaye, B., De Borman, B., Coumans, B. *et al.* (1999) Housekeeping genes as internal standards: use and limits. *Journal of Biotechnology* 75(2–3), 291–295. DOI: 10.1016/S0168-1656(99)00163-7.

Torres, T.T., Metta, M., Ottenwalder, B. and Schlotterer, C. (2007) Gene expression profiling by massively parallel sequencing. *Genome Research* 18(1), 172–177. DOI: 10.1101/gr.6984908.

Tzfadia, O., Diels, T., De Meyer, S., Vandepoele, K., Aharoni, A. *et al.* (2016) CoExpNetViz: Comparative Co-Expression Networks Construction and Visualization Tool. *Frontiers in Plant Science* 6(394), 329. DOI: 10.3389/fpls.2015.01194.

Untergasser, A., Cutcutache, I., Koressaar, T., Ye, J., Faircloth, B.C. *et al.* (2012) Primer3 – new capabilities and interfaces. *Nucleic Acids Research* 40(15), e115–e115. DOI: 10.1093/nar/gks596.

Vandesompele, J., De Preter, K., Pattyn, F., Poppe, B., Van Roy, N. *et al.* (2002) Accurate normalization of real-time quantitative RT-PCR data by geometric averaging of multiple internal control genes. *Genome Biology* 3(7), 1. DOI: 10.1186/gb-2002-3-7-research0034.

Wang, Y., Wang, Z.-K., Huang, Y., Liao, Y.-F., Yin, Y.-P. *et al.* (2014) Identification of suitable reference genes for gene expression studies by qRT-PCR in the blister beetle *Mylabris cichorii*. *Journal of Insect Science* 14(1), 94–94. DOI: 10.1093/jis/14.1.94.

Wilhelm, B.T. and Landry, J.-R. (2009) RNA-Seq – quantitative measurement of expression through massively parallel RNA-sequencing. *Methods* 48(3), 249–257. DOI: 10.1016/j.ymeth.2009.03.016.

Will, K.W., Attygalle, A. and Herath, K. (2000) New defensive chemical data for ground beetles (Coleoptera: Carabidae): interpretations in a phylogenetic framework. *Biological Journal of the Linnean Society of London* 71(3), 459–481. DOI: 10.1006/bijl.2000.0456.

Winnebeck, E.C., Millar, C.D. and Warman, G.R. (2010) Why does insect RNA look degraded? *Journal of Insect Science* 10(159), 159–7. DOI: 10.1673/031.010.14119.

Yu, N., Christiaens, O., Liu, J., Niu, J., Cappelle, K. *et al.* (2012) Delivery of dsRNA for RNAi in insects: an overview and future directions. *Insect Science* 20(1), 4–14. DOI: 10.1111/j.1744-7917.2012.01534.x.

Index

Page numbers in **bold** type refer to figures and tables.